The Cream of the Crop
Canadian Aircrew, 1939–1945

The Cream of the Crop is the first comprehensive account of the Royal Canadian Air Force's (RCAF) selection and training procedures and its policies governing aviators considered unsuited for operational duties, especially those judged to have a "lack of moral fibre" (LMF). Allan English assesses the effectiveness of RCAF use of manpower and questions whether the aircrew really were, as the RCAF alleged, the cream of the nation's crop.

English describes the development of a uniquely Canadian selection system that attempted to match aircrew candidates' aptitudes to the duties they would perform and the evolution of the RCAF's training program from a haphazard system with enormous attrition to one that became the model for many modern systems. He traces the evolution of aviation psychology and the treatment of psychological casualties of air combat. English pays particular attention to the LMF controversy and the RCAF's response as well as the effect of morale and leadership on the psychological well-being of, and casualty rates among, Royal Air Force and RCAF bomber squadrons.

In exploring the human dimension of air warfare, an issue that has been widely overlooked in military literature, English demonstrates that personnel considerations have at least as much influence on the effectiveness of air forces as material and technological factors.

ALLAN D. ENGLISH teaches in the war studies program at the Royal Military College of Canada. He served as an air navigator in the Canadian Forces from 1971 to 1991.

The Cream of the Crop

Canadian Aircrew, 1939–1945

ALLAN D. ENGLISH

McGill-Queen's University Press
Montreal & Kingston • London • Buffalo

© McGill-Queen's University Press 1996
ISBN 0-7735-1398-1

Legal deposit third quarter 1996
Bibliothèque nationale du Québec

Printed in Canada on acid-free paper

This book has been published with the help
of a grant from the Humanities and Social
Sciences Federation of Canada, using funds
provided by the Social Sciences and Humanities
Research Council of Canada. Funding has also
been received from the Department of History
at Queen's University.

McGill-Queen's University Press is grateful
to the Canada Council for support of its
publishing program.

Canadian Cataloguing in Publication Data

English, Allan D. (Allan Douglas), 1949–
    The Cream of the crop : Canadian aircrew, 1939–
    1945

Includes bibliographical references and index.
ISBN 0-7735-1398-1

1. Airmen – Training of – Canada – History.
2. Canada. Royal Canadian Air Force – Recruiting,
enlistment, etc. – World War, 1939–1945.
3. Aeronautics, Military – Canada – Psychology – History.
4. World War, 1939–1945 – Aerial operations, Canadian.
I. Title.

D792.C2E44 1996   358.4'15'0971   c96-900371-4

*This book is dedicated to two members of the Royal Canadian
Air Force who served their country proudly in war and in
peace, and inspired my interest in aviation.*

*To my parents,
James Stewart English
and
Jacqueline Andrée English*

Contents

Acknowledgments

In the more than six years I spent working on the doctorate that engendered this book, I was influenced by many people. The authors whose work has been useful to me are listed in the bibliography, but I would be remiss if I did not acknowledge some of the major personal influences on my ideas.

I am indebted to the staff of the War Studies Program at the Royal Military College of Canada (RMC), particularly its Chair, Ronald G. Haycock, Dean of Arts at RMC, and the late Barry D. Hunt, former Dean of Arts at RMC. The program's approach toward the study of human conflict affirms that war, as a focus of scholarly inquiry, is not a unidimensional field confined to military minutiae, but an all-encompassing subject that deals with social issues as well as diplomacy, strategy, tactics, and technology. It became clear to me during my time in the graduate program at RMC that war, aside from its more obvious effects, influences societies in many subtle ways as well. In many respects, the latter are every bit as important as the conspicuous results of war.

Another important influence on me was Donald Harman Akenson, who, as one of my thesis advisers, as teacher, and as mentor, showed me how history really works and how it should be done. I must also note the impact of the late Edward P. Thompson on my academic development. The graduate seminar he gave at Queen's University in the 1988 winter term taught me a great deal about the human face of history, and his many witty anecdotes and utter lack of dogmatism in his dealings with his students was a fine example of how one should teach history.

Two other scholars have also been instrumental in my academic career. Lieutenant-Colonel (retired) John A. English (no relation) encouraged me to persevere at an early stage in my doctoral studies, in spite of some obstacles that were being placed in my way at that time; he has remained a colleague and a source of good advice. Professor Emeritus Donald M. Schurman, who supervised my MA thesis and whose opinions I value in many matters, has become a friend and counsellor. A veteran of Bomber Command "ops" he was willing to share some of his experiences with me, and these have added greatly to the authenticity of the present account.

I am also grateful to those people who gave some of their valuable time to discuss the issues that interested me (if not necessarily them), to read chapter drafts, and to offer advice on how to improve this book. Besides those persons already named, this group includes Professor Terry Copp and Lieutenant-Colonel (retired) Karol Wenek (formerly of the Military Psychology and Leadership Department at RMC).

I particularly appreciate the advice and support given to me by historians at the Directorate of History, Department of National Defence. Carl Christie, Stephen Harris, Ben Greenhous, Bill McAndrew, Major John Armstrong, Lieutenant-Commander Richard Gimblett, and Major Jean Morin were all extremely helpful.

My research would have been impossible without the aid of the staffs at the Massey Library and the Science and Engineering Library at RMC, the Queen's University Libraries and Archives, and the Directorate of History. The staff of the National Archives of Canada was equally cooperative, especially Robert Miller of the Access Section, who helped me to secure the records I requested. The assistance of Sergeant L.W. Landry and his staff at the Canadian Forces Photographic Unit, Ottawa, was extremely helpful in finding the photographs for this book.

A project of this magnitude has left me indebted to many other people, and I am grateful to everyone who contributed to it even if I fail to acknowledge them here.

For financial support I wish to thank the Department of National Defence (Canada), which gave me Military and Strategic Studies PhD Scholarships for the academic years 1991–92 and 1992–93.

My family has sustained me throughout this experience, and the fruits of my labour belong as much to my wife Gina, my daughter Meredith, and my son Kyle as to me.

Abbreviations

CPA	Canadian Psychological Association
CSR	Combat Stress Reaction
DAOC-in-C	Deputy Air Officer Commanding-in-Chief
DDP1	Deputy Director of Personnel 1
DGMS	Director General of Medical Services
DHIST	Directorate of History (Department of National Defence, Ottawa)
DND	Department of National Defence (Ottawa)
DP	Director of Personnel
DPS	Director of Personal Services
EFTS	Elementary Flying Training School
FAI	Fédération aéronautique internationale
FIS	Flying Instructors' School
FPRC	Flying Personnel Research Committee
GOC	General Officer Commanding
HMSO	His (Her) Majesty's Stationary Office
HQ	Headquarters
ITS	Initial Training School
LMF	lack of moral fibre
MD	Militia District
MND	Minister of National Defence
MNDA	Minister of National Defence for Air
'MO	medical officer
NAC	National Archives of Canada
NCO	non-commissioned officer
NP	neuropsychiatric
NRC	National Research Council of Canada
NYDN	Not Yet Diagnosed Nervous/Neuropsychiatric
OTU	Operational Training Unit
PPO	Provisional Pilot Officer
PRO	Public Records Office
PS	Principal Secretary
QUA	Queen's University Archives
RAAF	Royal Australian Air Force
RAF	Royal Air Force
RAFSC	Royal Air Force Staff College
RAMC	Royal Army Medical Corps
RCAF	Royal Canadian Air Force
RCAMC	Royal Canadian Army Medical Corps
RFC	Royal Flying Corps
RNAS	Royal Naval Air Service
SFTS	Service Flying Training School
SMC	Secretary Militia Council

SMO	Senior Medical Officer
SOMS	Staff Officer Medical Services
USAAC	United States Army Air Corps
USAAF	United States Army Air Forces
US of S	Undersecretary of State for Air
VAD	Volunteer Aid Detachments
VLT	Visual Link Trainer
W	waverer

Royal Flying Corps (RFC) Canada cadets at the University of Toronto,
April 1917. (Department of National Defence [DND], RE 19,008-1)

RFC Canada cadets studying a JN4 aircraft at the School of Military
Aeronautics, University of Toronto. (DND, RE 19,062-40)

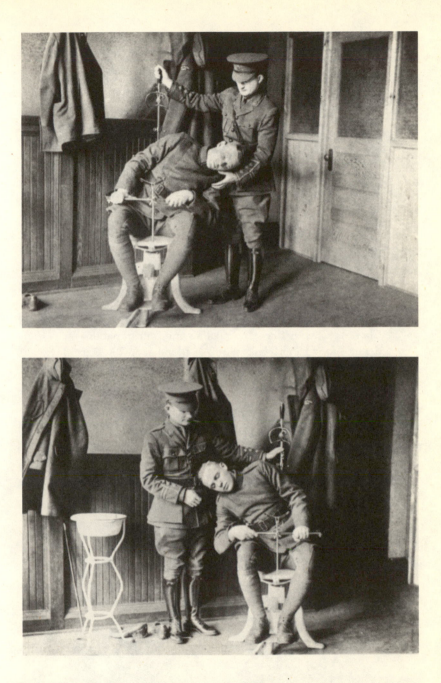

An RFC Canada recruit being tested in a rotating chair for "an acute and correct sense of equilibrium." (DND, RE 19,297-4)

Charles "Chubby" Power, Canada's minister of national
defence for air (left), talks with Air Marshal A.G.R. Garrod,
Air Member for Training, Royal Air Force (RAF), during a visit
to Britain, 4 July 1941. (DND, PL 4,412)

Air Marshal Philip Babington, Air Member for Personnel, RAF, left,
welcomes Air Vice-Marshal Harold Edwards to Britain as he arrives to
take up his post as Air Officer-in-Chief Royal Canadian Air Force (RCAF)
Overseas, 22 November 1941. (DND, PL 4,936)

Group Captain Sperrin Chant, the RCAF's Director of Personnel Selection and Research, in front of an RCAF Classification Test poster, 1 April 1944. (DND, PL 24,478)

Flying Officer D.C. Williams, PhD (right) receives his pilot's wings from Group Captain W.I. Clements at RCAF Station Uplands, Ottawa, 26 January 1945. As a young psychologist working for Chant, Williams conducted a scientific evaluation of the Visual Link Trainer as a pilot-selection device, and took pilot training as a "participant observer." (DND, PL 34,887)

Flying Officer A.E. Jarvis (right), one of the earliest Canadian proponents of the Visual Link Trainer as a pilot-selection device, instructing a student at No. 1 Initial Training School, Toronto, 25 July 1940. (DND, PL 1,051)

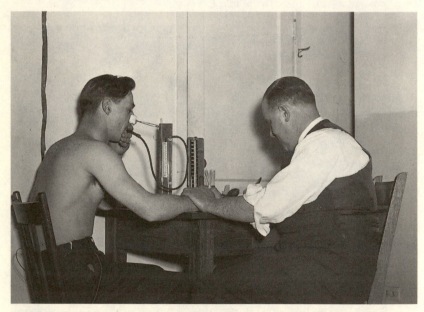

An RCAF aircrew candidate undergoing the "U tube test," March 1941. By mid-1941 both the RAF and the RCAF were beginning to question the usefulness of this World War I pilot-selection procedure, and it was abandoned by both services before the end of the war. (DND, PL 4,192)

RCAF applicants at the "approach desk" of a recruiting centre. (DND, PL 20,911)

Air force applicants writing the RCAF Classification Test at a recruiting centre. (DND, PL 20,913)

Three Canadian graduates of the British Commonwealth Air Training Plan at No. 1 Air Observer School, Malton, Ontario, June 1941. Left to right: Sergeants J. Nelson (pilot), B. Murray (observer), R. Abee (wireless operator (air gunner)). (DND, PL 3,633)

An RCAF Wellington crew in Britain, 19 May 1943. (DND, PL 19,009)

A crew from 420 Squadron before a mission, 21 May 1943. (DND, PL 15,978)

The effects of flying stress. Members of 417 squadron after a mission, 21 February 1943. (DND, PL 15,626)

A No. 4 Group Halifax on fire over Europe, 24 March 1945. Even if the crew survived such a harrowing experience, the flying stress caused by the event would usually have serious psychological consequences. (DND, PL 144,284)

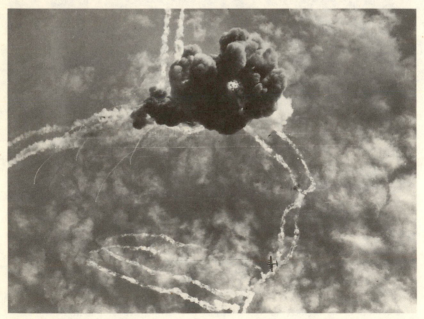

One of the reasons why so many Bomber Command casualties were listed as missing on operations. A No. 3 Group Lancaster explodes in mid-air, 19 February 1945. (DND, PL 144,292)

The Cream of the Crop

1 Introduction

One of the greatest problems a modern society can face in time of war is how to use the talents of its citizens most effectively. Invariably there are too few people to perform all the complex tasks required by a nation at war. This is especially true where flying and fighting in the air are concerned. Countries that choose to raise large air forces confront the difficult challenges of identifying people who are able, in a relatively short time, to learn how to fly, and, once the aviators are trained, of conserving those valuable resources as far as possible.

As early as the fourth century BC, the ancient Greeks speculated about how individual characteristics should be classified, and about how to identify the people most qualified to play various roles in society. Plato's judgment that it was necessary to choose leaders from among citizens endowed with "quick comprehension, a good memory, sagacity, acuteness," and trustworthiness has been revised very little in the intervening centuries.[1] What has changed is the scale on which modern states require citizens possessing the desired attributes, and the methods used to select, prepare, and care for such people.

World War I was the first example of a truly modern war, one which embodied the concept of the industrial nation-in-arms. The development of efficient government structures in the nineteenth century prepared the way for near "total war" in that conflict, which in the end demanded the mobilization of almost the entire material and human resources of the European belligerents. World War II made similar demands on the major combatants, but required better human material to operate the increasingly complex weapons of war.

For the military establishments of 1914–18 and 1939–45 to compete with their adversaries in an era of escalating violence, it was essential for them to use their personnel as efficiently as possible. This required the participation of specialists in the field of human behaviour – mostly civilians – to aid in the marshalling and conservation of each nation's fighting forces. The involvement of civilian experts in Canada's innovative approaches to the management of air force personnel in World War II is one part of the story to be told here.

The first efforts to bring expert knowledge to bear on problems of human behaviour were made in World War I, but the application of scientific principles on a large scale was only achieved in World War II. At the beginning of the twentieth century the science of human behaviour was still a relatively new discipline, but the use of increasingly sophisticated statistical techniques in the 1920s and 1930s allowed psychologists to assert themselves in the highly practical domain of identifying potential pilots and others with special abilities among the enormous populations of the nation-in-arms. However, once even carefully chosen and highly trained aviators entered combat they almost invariably suffered, in varying degrees, from the psychological stress of aerial warfare. A significant proportion experienced breakdowns and were lost to their air forces. Physicians trained to deal with abnormal human behaviour were called upon to minimize these losses and to salvage as many flyers as they could by developing special treatment protocols.

Military experts based their approaches to these problems on science as they knew it, but as Stephen Jay Gould has pointed out, science is not an entirely objective undertaking. It is also a social and cultural phenomenon that can effect change in its host society, and for this reason it is important to understand how it works in its social and cultural context.[2] For historians interested in the relations between science and society, periods of national upheaval, such as wartime, provide an excellent setting for analysis. In World War II the pressures of war forced the scientists who are the subject of this book to address problems that their military and political masters viewed as critical, and in so doing, the scientists' activities came under much greater scrutiny than would normally have been the case. Military organizations usually keep voluminous records, and the two great wars of the twentieth century bequeathed to Canada a public archival record that enables us to study science in action at a level of detail usually denied to researchers who look at the endeavours of scientists in civilian settings.

World Wars I and II were pivotal events in the history of Canada, and they were vital in "defining our national sense of self."[3] In 1914

Canadian troops were part of a large imperial army. They were fully integrated into the British forces, and the Canadian government exercised little control over their destiny. After four years of bloody sacrifice, Canada emerged as a "junior but sovereign all[y]" in the western coalition.[4] Twenty years later, Canada participated in global conflict as a fully autonomous state,[5] and despite its small size relative to the major powers, it made important contributions to the Allied struggle. In World War II Canada boasted the third-largest force of fighting ships in the world and the fourth-largest air force, and stood fourth in importance as a supplier of materiel to the Allies.[6] This was a remarkable accomplishment for a country that before the war "could barely eke out enough political stability and economic strength to preserve a precarious and unsatisfactory existence in North America, without indulging in any European adventures."[7] Canada's impressive feat was made possible by achievements in many areas, but of fundamental importance was the way it organized the people who produced and operated the machinery of war. As a matter of policy, a large number of them were employed in one of this nation's most important contributions to the conflict – the Allied air effort.

Soon after World War II began, Canada embarked on a massive aircrew-training program known as the British Commonwealth Air Training Plan (BCATP), which was intended to transform the country into the "aerodrome of democracy."[8] The Royal Canadian Air Force (RCAF) was responsible for implementing the BCATP, which trained more than 130,000 aircrew of virtually every Allied nationality. During the war over 250,000 people served in the RCAF, or almost one in four of the men and women recruited by Canadian armed forces. Charles "Chubby" Power, the minister of national defence for air, remarked in November 1941 that the RCAF had "taken the very cream of the youth of Canada ... They are the future leaders of this country, and the destiny of Canada will some day be in their hands."[9]

Tragically, many did not live to see this destiny fulfilled. The carnage among Commonwealth aircrew over European skies was so high, in fact, that it has been equated to the "slaughter of the [British] Empire's élite" in World War I.[10] The RCAF's share of the Canadian sacrifice in World War II was 17,101 dead, representing about 40 per cent of the fatal casualties suffered by the Canadian forces. What is remarkable is not only that the RCAF paid a very high price compared to its total strength, but also that 92 per cent of all air force casualties were fatal, compared to 30 per cent for the Canadian Army. Furthermore, while ground staff outnumbered aircrew five to one in the air force, aviators accounted for 94 per cent of the dead.[11]

These awful losses among such a carefully selected and trained group of people had important implications for the nation, and should also provide lessons for future conflicts. With this in mind, the present study examines three issues vital to Canadian manpower[12] policy in World War II – aircrew selection, aircrew training, and the procedures governing the employment of aircrew, especially as they were affected by attitudes toward so-called lack of moral fibre (LMF). In the same way that what Marc Milner has called the "new military history" seeks to broaden its frontiers by exploring "the links between the military experience and the larger society,"[13] the present study highlights the relationship between Canada's evolution as a sovereign state and the air force's manpower situation.

As Richard Ruggle has observed, most studies in military history have had a "preoccupation with weapons, tactics, technology and grand strategy," but have generally "ignored the vast human dimension of battle."[14] Yet as Martin van Creveld has pointed out, to understand warfare it is necessary to comprehend "the *totality* of factors involved,"[15] including how a nation's human resources are organized, trained, and deployed. Despite the ample attention paid by students of warfare to the machines used in battle, technology "on its own will seldom decide a war." The relationship between people and machines makes technical systems more than just a collection of equipment; they become complete philosophical systems as well.[16] The key to success in any conflict is to recognize the limitations of technology and to overcome them with "improvements in training, doctrine, and organization."[17]

This was first recognized in World War I, when many states underwent a "managerial metamorphosis"; old ways of doing things were swept away and "technocratic élites" were called upon to organize national resources on an unprecedented scale by applying rational criteria to state planning. According to William McNeill, the chief lesson of World War I was that "manpower, food, and fuel were the ultimate regulators of ... war effort." To deal with these constraints war began, of necessity, to become civilianized. In 1939–45 the combatants sought to overcome limitations on national resources by organizing on a "[t]ransnational" basis, which led to a "more or less rational and deliberate division of labour" between Britain and the United States.[18] At the same time, large numbers of civilian scientists were incorporated into all levels of government and into most military headquarters.[19] By virtue of their privileged positions close to political and military leaders, certain scientists and other specialists had an opportunity to mould some of those societal values that were precipitated into a state of flux by the stresses of war.

This book focuses on the wartime roles of two communities of civilian specialists – Canadian psychologists and British physicians – in the Commonwealth air effort. Canadian psychologists were among the leaders in the field of aircrew selection by the end of World War II, but in 1939 there was some doubt that they were qualified to play any part at all in the country's mobilization effort. By the start of the war, researchers in the behavioural sciences were strongly influenced by the concept of "human engineering." The proponents of this idea held that workers' tasks should be broken down into their basic elements, which could then be controlled by the workers' supervisors. The RCAF's psychologists conceived its earliest selection methods based on this concept, and they attempted to measure very simple, discrete abilities, such as reaction time to a stimulus, in the hope of discovering some elementary way of identifying potential aircrew. However, the men who created the Canadian aircrew-selection system, some of the most prominent civilian psychologists in the land, soon recognized the futility of this approach, and eventually devised an original selection system based on practical and relatively complex tests that imitated some aspects of flying an aircraft. Mass testing and the mathematical manipulation of data, adapted to psychological use in the United States during World War I, was the model used by RCAF psychologists to process large numbers of aircrew candidates in a relatively expeditious and efficient manner.

Unlike the mechanical and rather straightforward selection process experienced by most Canadian aircrew, those who required treatment for psychological disorders or "flying stress" were dealt with by specialists who used more subjective methods of handling these problems. In the Commonwealth air forces engaged in the fight against the Axis, the mental health of aircrew was largely the responsibility of civilians in uniform, namely psychiatrists and other members of the British medical profession. Unlike psychologists, of course, psychiatrists are qualified physicians, and their diagnoses and treatment regimes are based on clinical judgments formed by extensive interviews with the patient. For them, normality is in large part a social construct, and psychiatric consensus about what constitutes a "normal" person is chiefly a social statement that varies over time and between cultures.[20] As we shall see in this book, psychiatrists helped to define normality among aircrews. In addition, by contributing to the definition of social norms in the air forces, and by assisting in the formulation of policies that moderated aircrew wastage rates, psychiatrists had a significant effect on aircrew attrition. But despite their contributions to the treatment of airmen with psychological problems, their demands to participate in the aircrew-selection system were rebuffed. Critics of the psychiatric approach complained that time-consuming psychiatric assessments were impractical when "total war" demanded rapid recruiting of great numbers of fighting men.

As scientific activities, psychology and psychiatry shared certain assumptions, and both pursued the truth as they perceived it. In addition, each discipline cultivated traditions that guided its members in selecting methods of inquiry and in establishing standards that were used to measure success in each field.[21] The methods chosen by each discipline determined, to some extent, the questions to be asked and the answers that were found. Each specialty, moreover, influenced the norms of a wartime society. For example, most aviation psychiatrists were guided by the assumption that the complexities of human behaviour could only be explained by a detailed understanding of the unconscious mind, whereas many aviation psychologists believed that only discrete measures of conscious behaviour, such as reaction time, were valid predictors of performance.

While psychology and psychiatry contributed to the shaping of social values in wartime, the demands of war, by the same token, altered psychological and psychiatric practices. Military and government leaders and the bureaucratic apparatus of Canada and Britain imposed new goals on scientific activity, furnished new criteria for evaluating what was relevant in research, and dispensed new rewards to the experts. The nation-in-arms asked its behavioural scientists to make as many of its citizens as possible available for service to the state, and for a variety of reasons the experts were anxious to oblige.

Although often overlooked in the historical literature because of the apparent dependence of air forces on technology, the optimal use of human assets is just as important in fighting an air war as it is in fighting on land or at sea. The official history of the British bombing offensive against Germany has concluded that the two main variables that limited the sustainable loss rate were "operational efficiency and morale" – both primarily human questions.[22] We should not downplay the importance of technology, but it must be recognized that personnel considerations have at least as much influence on the effectiveness of military organizations as do factors related to materiel.

As Stephen Harris and Norman Hilmer have noted, no "comprehensive account of air force training or the way in which the RAF [Royal Air Force] managed its personnel resources" has been written, despite the importance of those subjects. Moreover, "questions about the composition of the air force, and personnel selection processes, are largely unanswered, while those relating to morale and discipline continue to be matters of conjecture."[23] Not only has there been little historical inquiry into military manpower issues generally, but until recently the subject of psychological casualties also was largely neglected. Terry Copp and Bill McAndrew's groundbreaking work has made a start on the study of "battle exhaustion" in the Canadian Army in World War II, but nothing of substance has been written about battle exhaustion in the RCAF, or in other air forces for that matter. Yet battle exhaustion has a long history. As Douglas Robinson has observed, "Perhaps the most serious failure of aviation medicine in World War I was the permanent loss from flying duty of many trained airman due to flying fatigue and/or war neurosis."[24] Battle exhaustion was also a problem for the air forces of World War II, and was one of the most serious causes of manpower loss in that conflict.

Statistics on the numbers of Canadians under arms in World Wars I and II offer some insight into the daunting challenges to the efficient management of human resources confronted by Canada in 1914–18 and 1939–45. In World War I, 640,000 Canadians were in uniform, or about eight per cent of the national population,[25] and 400,000 went overseas. By the end of the war one-quarter of Canada's men of military age had been killed or wounded.[26] In World War II, over one million Canadians joined the fighting services, or about nine per cent of the national population. More than 40 per cent of the eligible male population were in uniform, and 96,456 Canadians were wounded or died on active service.[27] The size and the nature of such manpower commitments caused conscription crises in both conflicts that threatened to tear Canada apart, and one of the major criticisms of the governments of the day was that they did not make the best use of available manpower.

In the selection of military personnel, the ideal method for efficiently employing a nation's human resources involves matching individual talents to the necessary jobs, but reallocating civilian manpower for military purposes has always been a process of trial and error.[28] In the history of warfare, most "armies seem to have prepared their campaigns as best they could on an *ad hoc* basis," gathering together the largest possible collection of assets,[29] including human beings. This was precisely the course Canada took at the beginning of both world wars. As we shall see in chapter 2, significant progress was made in matching aircrew candidates' aptitudes to the duties they were required to perform. Chapter 2 also describes the development of a uniquely Canadian aircrew-selection system, which provides the basis for the system used in Canada today.

Chapter 3 looks at the training of aircrew, particularly pilots, and the progress from a haphazard system with enormous attrition to one that became the model for many modern training schemes. A great many lessons were learned in Canada and in Britain during World War I about the effective instruction of aviators, but very few of them were put into practice until late in the next war, and avoidable manpower wastage resulted.

Psychological casualties represented one of the greatest sources of aircrew attrition in both world wars, and chapters 4 through 6 trace the evolution of procedures for their treatment. This book devotes more space to psychological casualties than to any other topic because, compared to training and selection, it is the least-studied subject in the literature. Chapter 4 describes the birth of aviation psychology in World War I, and how that discipline was transformed from 1918 to 1945. Chapter 5 examines the LMF controversy, and how morale and leadership affected the psychological well-being of, and casualty rates among, RAF and RCAF bomber squadrons. Chapter 6 details how the RCAF grappled with the LMF issue, which also included the larger question of how to handle aircrew who were unable or unwilling to perform their duties. It also examines the methods Canadians eventually created for dealing with aviators suspended from flying duties.

Chapter 7 discusses the impact of RCAF manpower policies on the two Canadian conscription crises of World War II and evaluates the effectiveness of RCAF manpower use.

Chapter 8, the concluding chapter, ties together the themes of aircrew selection, aircrew training, and the use of human assets in war. It summarizes the lessons learned in the previous chapters, and also assesses the nature and the role of institutional memory in military organizations. Finally, it asks whether RCAF aircrew really were Canada's "cream of the crop."

The remainder of the present chapter offers a brief sketch of the history of the Canadian air force from its beginnings in World War I to its maturity in the next war, and considers its relationship to national political and military policy. The story has been told elsewhere, but its recounting here will provide a context for the discussion that begins in chapter 2.

The Royal Flying Corps (RFC) and the Royal Naval Air Service (RNAS), the air arms of the British army and navy, were combined to form the RAF in April 1918, although a number of organizational changes (such as the formation of an Air Council in January) preceded the formal birth of a unified British air force.[30] In June 1918 Canadian and British officials agreed to create a small Canadian Air Force (CAF),[31] which was granted the honorific "Royal" in April 1924 in recognition of the "'efficiency, gallantry and devotion to duty'" of the "thousands of Canadians who had flown with the British flying services during the war."[32]

Canada took great strides toward nationhood in World War I, but in the air the gains were more tentative than elsewhere. While the Dominion supplied many of the young men who fought in the RFC/RAF, the British flying service was organized as an imperial force. But despite its imperial designation, perhaps 40 per cent of the aviators in the RAF on the Western Front at the end of World War I were Canadian.[33]

After the war there were some 13,000 trained aviators in Canada, but the rush to demobilize and to forget things military left the country with an air force that comprised "bush pilots in uniform."[34] The bush pilots opened the north for development, and asserted Canadian sovereignty over much of the territory to which the country laid claim, but concentration on civilian tasks left the CAF/RCAF subordinate to the Canadian militia and only nominally a separate service. In 1937 the RCAF was granted autonomy, but until the war years it remained subject to a "ramshackle administrative system."[35] This was to have serious repercussions for its wartime expansion. The economies of the interwar years left the CAF/RCAF without any staff who could maintain or develop expertise in important areas outside of routine flying operations. Also, because almost all its professional development courses were conducted in Britain, many future senior Canadian air force officers were exposed mainly to the RAF way of doing things.[36] This helps to explain why early RCAF personnel policies were often copies of the policies followed by the RAF. However, the methods propounded by the RAF in peace were not always appropriate to war. According to the official history of the Canadian air force, one of the reasons for deficiencies in the performance of the Canadian air staff in World War II may have been

the fact that the officers who rose to senior rank had not been properly prepared to organize, control, supply, and direct a large air force. There is always a danger when minuscule professional forces are compelled to expand quickly in wartime: no matter how earnest, hard-working, and determined the air staff officers may have been, their peacetime experience bore no relation to the demands made of them once the war began. Little wonder, then, that Air Force Headquarters at times seemed out of its depth.[37]

In the late 1930s the RCAF began a gradual expansion, but even when it finally gained its administrative independence from the militia in November 1937 it remained tiny. One officer recalled that "'On the eve of World War II there were only 290 officers and 2,700 other ranks in the permanent RCAF. There were so few of us that we knew one another either personally or by reputation.'" This intimacy persisted at senior levels even with the air force's rapid expansion.[38]

Notwithstanding the neglect of the air force between the wars, when another European conflict appeared inevitable in the 1930s, the RCAF was chosen by the Canadian government to be a key instrument of the nation's military strategy. In 1937 the prime minister, William Lyon Mackenzie King, designed his defence policy to "appeal to moderate opinion," and, because of the "growing influence of air-power," one built on expanding the size of the RCAF.[39] However, even though Mackenzie King's minister of defence, Ian Mackenzie, announced in Parliament in April 1939 that the "'first line of defence for the Dominion of Canada must be the Air Force,'" and allocated one-half of the $60 million Department of National Defence budget to the RCAF, by the outbreak of war in September the RCAF was manned to only just over one-half of its authorized establishment of 7,259.[40]

But perhaps a slow build up of military force was the only policy that could have kept the country together. If one accepts the observation of another prime minister, R.B. Bennett, that Canada is "'one of the most difficult countries to govern,'"[41] then Mackenzie King's ability to bring a united nation into World War II can be regarded as a "remarkable and improbable achievement."[42] As a writer on the mobilization question remarked in 1941, before "September 1939 the temper of the Canadian people was pacific ... [and] 'appeasement' was a popular policy." The same writer reflected the prime minister's feelings when he expressed his concern about Canada being "dragged into a European struggle every twenty years."[43] On the financial front, one of Mackenzie King's closest advisers, O.D. Skelton, warned that "Canada was having a hard time paying for the First World War; how could it even begin to pay for a Second?" The new war "threatened political and racial division ... and financial bankruptcy." In the face of such criticisms, Mackenzie King initially adopted what might be called a policy of "rigorous restraint," but as time went on the Canadian government became increasingly committed to the Allied war effort.[44] Throughout the conflict manpower remained "Canada's most durable military problem," and conscription dominated the nation's politics.[45]

Thus the decision was reached at the highest political levels that Canada's main contribution to the war would be in aviation, and that casualties would be kept low by confining Canadian air force activity to home soil. This plan was supported by the British, who believed that the "supreme effort was to be made in the air."[46] With the signing of the BCATP in December 1939, it appeared that the perfect answer to limiting the Canadian manpower commitment had been found. In the early months of the war, however, a few Canadian squadrons were sent overseas "'to satisfy public sentiment,'"[47] because "'Canada's prestige as a nation'" demanded it, and because it was believed that "Canadians by temperament would 'prefer to be at the front' in Canadian units."[48]

But in formulating wartime policy, the Canadian government overlooked one crucial fact: a vast domestic aircrew-training scheme was sure to attract large numbers of the Dominion's youth. In an era when airplane travel was restricted to a privileged few, the opportunity to learn how to fly and to fight in the air must have been incredibly tempting for many young men. At first, the BCATP produced more aircrew than overseas squadrons could absorb, but the longer the war lasted the greater was the number of Canadians who were sent to fight in the skies over Europe. The experience of 1914–18 had shown that casualties among airmen could be extremely high, and that long-range bombing (the RAF's preferred method of waging air warfare in World War II) had resulted in particularly high losses among the attackers. The Germans had been forced to call off their attacks on London owing to attrition in their bomber forces, and the shorter-range Allied efforts against Germany had also suffered severely.[49] But these lessons were not recalled. For the moment, the plan seemed perfect for Canadian needs, and, in the end, it did accomplish a great deal. The largest aircrew-training organization in the British Empire, and arguably Canada's "greatest achievement" since Confederation,[50] the BCATP furnished 44 per cent of the 340,000 Commonwealth aircrew trained from 1939 to 1945.[51] However, the success of the plan in producing huge numbers of aircrew should not blind us to the fact that the vast majority of its graduates were assigned to fight in ways and in places largely determined by other nations.[52]

Before World War II began there were as few as four hundred Canadians serving in the RAF,[53] but in the later 1930s, when a second conflict with Germany seemed imminent, Britain began to look to "its imperial partners for manpower and the training space" needed to field a wartime air force.[54] There was "a clear precedent" for the BCATP in the imperial organization known as RFC (later RAF) Canada, which had trained "two thirds of the 21,000 Canadians who served in Britain's air forces in the First World War." But that war had also made it clear that Britain preferred individual airmen for imperial service, not formed Dominion squadrons. This view was bluntly expressed by a senior RAF officer who was quoted as saying that the RAF simply wanted "'bodies.'"[55] But British attitudes conflicted with Canadian "national sentiment," and despite the best efforts of the mother country to recruit Canadian bodies, the British eventually concluded that "'the political factor was too strong.'" Consequently, as the war progressed, more and more Canadian flyers were sent overseas in RCAF squadrons, not as unaffiliated individuals.[56] By 1945, forty-eight of the RCAF's eighty-eight squadrons had been dispatched abroad.[57]

The fighting force that the RCAF contributed to more than any other was RAF Bomber Command. In the dark days of late 1940 Winston Churchill could see "'only one sure path'" to victory; with no continental army left in the field to engage Hitler, "'an absolutely devastating, exterminating attack by very heavy bombers ... upon the Nazi homeland'" was believed to be Britain's only hope of changing the course of the war.[58] For proponents of the knock-out blow from the air there were "no limits on the size of the force required." In October 1942 the British Chiefs of Staff Committee accorded the bomber fleet "absolute priority of Anglo-American production" until 1943,[59] and it grew to become the RAF's largest component, encompassing one-fifth of RAF personnel.[60] From Dunkirk until well after D-Day, Bomber Command was the largest Allied force capable of attacking the German heartland.[61]

Churchill's belief that the war could be won – and heavy British casualties on the ground avoided – through strategic bombing led him to favour Air Chief Marshal Sir Arthur Harris, Commander-in-Chief (C-in-C) of Bomber Command, over any other military leader. Harris received frequent invitations to visit the prime minister at Chequers, his country retreat, where operational matters were discussed outside of the chain of command.[62] The official history of the bombing campaign has claimed that "Never ... in British history had such an important Commander-in-Chief been so continuously close to the centre of government power as Sir Arthur Harris was to Mr. Churchill."[63] And as one historian has remarked, this close personal relationship "gave Harris the power to bully and often to dictate to the Air Ministry," and allowed Harris to appropriate "the maximum amount of national effort in man-power and material for his Command."[64]

While the exploits of "the few" in such epic struggles as the Battle of Britain are better known among the general public, Bomber Command recruited the lion's share of RAF and Canadian aircrew. In

January 1945, 46 per cent of Bomber Command's pilots came from Canada, Australia, or New Zealand, and of this group 55 per cent were Canadian.[65] Bomber Command lost more men, and suffered the highest number of psychological casualties, of any branch of the RAF, and the only limit to the number of young Canadians sent to serve in it was the speed with which the training system could turn them out. By the end of the war 47,268 men had died on Bomber Command operations – more than two-thirds of all deaths in the RAF in World War II.[66] RCAF aircrew accounted for 9,919 of those deaths,[67] and more than one-half of all RCAF personnel killed in World War II belonged to Bomber Command. Of the total number of Canadian military deaths in 1939–45, the men under Sir Arthur Harris's leadership accounted for about one-fifth.

Sheer numbers alone, then, guarantee Bomber Command a prominent place in the chapters that follow. Another important reason for focusing on Bomber Command is the enormous sums of money that were spent on the force. It was estimated, for example, that the cost of training just one member of a Bomber Command aircrew was £10,000, "'enough to send ten men to Oxford or Cambridge for three years.'"[68] According to Sir John Slessor, this expense seems to have been justified, as the men of Bomber Command epitomized one of the "'two ... most unbelievable manifestations of human courage and endurance in the history of war,'" the other being "'the infantry of 1914–1918.'"[69] Even though operational tour lengths were designed to give aircrew a 50–50 chance of survival, losses were often higher, and during certain phases of the bomber offensive as many as 75 per cent of the crews perished.[70]

The BCATP and Canada's sacrifices in the European bombing campaign eventually helped to transform the RCAF into a national air force, but in many respects it lagged behind the Canadian Army in establishing its own identity. Just as the Canadian Army of 1914 was placed under British direction, the RCAF began World War II "leaving the allocation of individuals after they had been trained altogether to the discretion and direction of the Royal Air Force,"[71] which, for operational and administrative convenience, preferred crews of mixed nationality under RAF command. C.P. Stacey contended that this practice, and the way the BCATP "was set up in 1939" to emphasize training roles for Canadian airmen, deprived senior RCAF officers of operational experience. He asserted that wartime manpower policies "'broke the back' of the RCAF and militated against its putting in the field a national air force comparable to the First Canadian Army."[72]

The Canadian authorities made efforts to remove Canadian airmen from RAF control by increasing the number of squadrons and larger formations under RCAF command. Finally, in January 1943, Canada established its own bomber group. According to William Carter, this gave the RCAF "the same separate status as the Canadian Army, namely British operational control coupled with Canadian administrative control." Carter has suggested that this could have been achieved in 1939 if the Mackenzie King government had opted for "a separate air force overseas with the same status as the Canadian Army."[73] However, it may be that the challenges facing the RCAF in 1939 were great enough without Canada taking on the added burden of establishing an autonomous air force overseas. We will begin to examine these challenges in the following chapter, which looks at the initial stage in the production of Canadian military aviators – aircrew selection.

2 "A Good Airman Is Worth a Battalion of Infantry":[1] Aviator Recruiting and Selection, 1914–45

Written procedures for the selection of combatants date from at least biblical times, when the people of Israel recorded practical instructions for choosing who would fight in time of war.[2] One of the qualities the ancient Israelites examined in choosing their warriors was courage:

And the officers shall speak further unto the people, and they shall say, What man *is there that is* fearful and fainthearted? let him go and return unto his house, lest his brethren's heart faint as well as his heart.[3]

Ideally, a good selection system should employ tests that are as simple as possible and that have demonstrable value in identifying people with the characteristics desired by those doing the selecting. However, until technocratic élites, especially psychologists, became involved with military selection generally and with aircrew selection in particular, little had changed from biblical times. Specialists began to take a serious interest in the selection process in World War I, when the United States Army, at the instigation of the American Psychological Association, created the most systematic military selection program yet seen. Under the Association's guidance the Army administered what were called the Alpha and Beta Tests to over 1,700,000 men. The tests were designed to measure intelligence as a basis for identifying men with aptitudes relevant to military occupations. On the basis of these tests soldiers were assigned to specialized courses of instruction, such as pilot training. Military psychologists today generally accept the view of the Chief Psychologist of the United States Army that these tests "legitimized" the participation of psychologists in military selection and "established important precedents" for the future.[4]

However, these same tests have been disparaged by Stephen Jay Gould. He has noted that American psychologists, in an attempt to transform psychology from a "soft" science into one as rigorous as physics, exploited mental testing in wartime as a way of gaining the acceptance of, and funding from, government authorities. Gould has also pointed out that these scientists, most of whom were convinced that ability was determined by heredity, relied on unproven assumptions about human behaviour to produce tests that reflected social and cultural biases, and ignored many of the complex issues surrounding the measurement of ability. The simplest of these tests rarely demonstrated any validity in identifying people with the special talents required by various branches of the American armed services. Using data from the United States Army tests, the psychologists moreover, arrived at certain conclusions that were later shown to be false, for example that 37 per cent of white Americans were "morons."[5]

Few practising military psychologists have responded publicly to Gould's criticisms; they prefer to portray their discipline as one driven by an improved methodology along a road of continuous progress. This is particularly true of aviation psychologists, and whatever debate there is over the validity of psychological criteria in aircrew selection, it remains on the periphery of their professional interests.[6]

Miliary psychologists who work in the area of selection draw on principles developed during World Wars I and II, and have adopted the concept of "criterion-related validity" as the cornerstone of their efforts. Criterion-related validity is based on measuring the differences between test results and performance in training or on operational duties. If there is a strong positive correlation between the two, the tests are considered valid.[7] It is rare for military organizations to allow those who fail selection tests to move on to training, especially flying training, because of the costs and risks involved. Therefore, there is always some doubt about how those not selected for training would have fared had they been allowed to undergo training.

The saga of Canadian aircrew selection in World War II encompasses most of these issues, and offers one of the few examples in the history of military personnel selection where relatively large numbers of individuals who did not meet selection standards were permitted to train for war. The Canadian experience showed that increasingly complex tests, more closely approximating the job itself, were required to predict with any accuracy who would and who would not successfully complete aircrew training within the time limits imposed by the military situation.

While the essentials of Canada's aircrew recruiting and selection arrangements have been briefly described in the official history of the Canadian air force, a comprehensive account has yet to be written. It is not this book's purpose to provide such an account, but it will trace the development of aviator selection in this country from its roots in World War I to its fruition in World War II. The focus will be on the role, during both world wars, of two civilian professional communities, the medical and the psychological, which were instrumental in the evolution of a distinctive recruiting and selection organization. This organization reached its apogee in World War II, when it chose for aircrew training only about 90,000 of the 600,000 men who applied.[8]

Canadian specialists first became involved in aviator selection during World War I, and as in Canada's air effort generally in that conflict, they for the most part imitated British models and developed a separate professional identity only late in the war. As was mentioned in the previous chapter, Canada had no national air force in World War I, and its citizens could only serve the British Empire as military flyers by joining the RAF or its predecessors the RFC and the RNAS.[9] For this reason, RFC/RAF selection procedures will be examined here in some detail.

The earliest applicants to the RFC were given the same medical examination as other members of the British Army. Men who transferred into the RFC from other units were examined by their own medical officers (MOs) or by a Royal Army Medical Corps (RAMC) board, while "new entrants were examined by the usual recruiting boards."[10] These selection agencies lacked familiarity with the physical requirements of flyers, and the result was a great deal of manpower wastage. Infantry standards permitted some aviation candidates to be accepted when almost blind in one eye, and therefore not fit to fly, while others were rejected solely because of flat feet, which were no handicap to a career as an aviator. In addition, because the army looked on flying duties as a way of resting soldiers who had been worn out or wounded while fighting in the trenches, a number of men suffering from battle exhaustion or with severe physical disabilities were admitted to the RFC, where most quickly broke down again or were unable to fully perform their duties.[11]

RFC recruiting problems were, in part, a consequence of the enormous strain the war imposed on the medical branch of the army.

At the outbreak of the war, the medical services in both Canada and Britain were overwhelmed by a flood of volunteers. The RAMC, for example, was prepared to screen 50,000 volunteers annually, but in September 1914 500,000 presented themselves for enlistment. As a result, some volunteers received only cursory medical scrutiny, or none at all. Continual changes in both countries in the rules governing military service and in medical standards, as well as a chronic shortage of qualified MOs, exacerbated the situation. From 1914 to 1918 the British and Dominion medical services treated 11,000,000 casualties; in the face of such an overwhelming burden on medical staff, efforts to remedy the problem of inadequate medical screening of recruits faltered.[12]

In some instances the actions of people in positions of high authority aggravated an already bad situation. Sam Hughes, Canada's minister of militia and defence, established a decentralized recruiting process that left the professional medical staff at headquarters with little to do. Those doctors promptly embarked for France, leaving "the examination of recruits ... largely ... in the hands of local practitioners who were without the advantage of experienced control." By January 1915 the pool of recruits was beginning to dry up, and commanding officers were able to choose "their own medical examiners who were governed by personal direction rather than by established standards." In effect, this enabled units to create their own selection criteria to fill their quotas.[13] Another obstacle to efficiency was the bitter quarrel between regular army doctors and Hughes over his intervention in the administration of the Canadian Army Medical Corps (CAMC). The intense infighting afflicted the entire medical service,[14] and adversely affected air force recruiting when it began in earnest in 1917.

Before 1916, when the RFC suffered disastrous losses on the Somme, Britain had little interest in obtaining Canadian aviators. However, with the need for more airmen, the British developed schemes to tap the Canadian manpower pool. The end of 1916 saw the establishment of RFC Canada, an autonomous imperial organization assisted by the Canadian government.[15]

The first major aviator recruiting campaign in Canadian history began in May 1917, and at the end of that year approximately 2,400 cadet pilots were enrolled in RFC Canada. Candidates were recommended by committees of local citizens and examined by the militia department's standing medical boards.[16] Problems arose, however, because Canadian Expeditionary Force (CEF) enrolment standards were "entirely different" from those of RFC Canada. The CEF criteria emphasized fitness for the infantry, whereas the RFC, whose standards were "much stiffer than [those of] any other branch of the service," required excellent vision in both eyes, and an above-average sense of balance. In addition, there were too many militia medical boards, and they were staffed by large numbers of unqualified examiners, so that candidates received inadequate examinations. These difficulties, and the problems already mentioned in connection with the CAMC, compelled RFC Canada to centralize its final acceptance procedure at its Toronto headquarters.[17]

The doctors who directed aviator selection assumed that flyers were men with special characteristics that could be identified by a detailed physical examination. Canadian MOs first received instructions regarding the examination of aviation candidates in March 1917, at RFC Canada's depot on the Exhibition Grounds, Toronto, when they were provided with RFC fitness standards. This marked the beginning of a systematic approach toward the medical examination of aviation candidates in Canada.[18] Prospective recruits were required to have perfect vision, to be under twenty-five years of age, and to be able to hear a whisper at twenty feet.[19] Acceptance was also conditional on the heart and blood vessels being "absolutely sound." Men with astigmatism, epilepsy, asthma, a history of nervous breakdowns, or a tendency to suffer respiratory infections were summarily rejected. In late 1917 the age limit was increased to thirty, the maximum acceptable weight of a recruit was fixed at 175 pounds, and a colour vision test was introduced. In July 1918 the age qualification was tightened when it was decreed that men twenty-six to thirty years of age would be accepted only "in special cases."[20]

Among the many duties performed by MOs, the examination of recruits was high on the list of priorities, and usually about one-third of RFC Canada's MOs were assigned to recruiting duties.[21] By November 1917, twelve MOs had been attached from the CAMC to RFC Canada, but the number of MOs assigned to RFC/RAF Canada varied from a low of eight in March 1918 to a high of thirty-two in September of that year.[22] To establish his authority over the MOs and to insulate them from the infighting in the CAMC, Brigadier-General C.G. Hoare, General Officer Commanding (GOC) of RFC Canada, asked in December 1917 that his medical service be "disassociated" from the Militia Districts except for supplies. Hoare's request was granted in February 1918, and Captain (later Major) Brefney O'Reilly was retained as the Senior Medical Officer (SMO) for RFC Canada.[23] Under the new regime the MOs continued to perform general medical duties, such as dealing with outbreaks of scarlet fever and influenza, but they were now able to devote more time to the practice of aviation medicine, including the treatment of psychological problems unique to aviators. It appears that most MOs attached to RFC/RAF Canada demonstrated a keen interest in the new field of aviation medicine. Their SMO encouraged them to get to know their charges well and to "become the confidant and adviser of all ranks." He also urged them to become acquainted with "the physical phenomena of flying" by getting "into the air as soon as feasible."[24]

One subject of great interest to physicians was the pilot's sense of balance. Many researchers believed that pilots, like birds, needed a good sense of balance to maintain their equilibrium in the air, and this sense of balance was thought to depend mainly on the semicircular canals of the inner ear.[25] This resulted in a requirement, peculiar to the flying service, of "an acute and correct sense of equilibrium" in all aviators.[26] A number of investigations, ranging from simple motor coordination exercises to tests conducted in rotating chairs, were used to assess this attribute. The tests employed in Canada eventually conformed very closely to those used by the United States Army Air Corps (USAAC).

A contemporary observer claimed that the American tests were adopted in Canada only after careful analysis of the results of both those students who had succeeded in and those who had failed their flying courses.[27] Actually, the tests had no value in predicting success in flying training, because pilots maintain their orientation in the air largely by visual reference to the horizon, and the organs of balance located in the inner ear are a secondary, and often misleading, source of information. O'Reilly himself was involved in experiments that helped to discredit the equilibrium tests; unfortunately, before the tests' alleged validity could be disproved, their stringent application deprived the RFC of many suitable candidates at a time when pilots were urgently needed.[28]

But medical tests were not the only hurdle aviation candidates had to clear. In 1917 and 1918 the first was an appearance before one of the civilian committees that formed part of a national pilot-recruiting system established by the Aero Club of Canada.[29] Created in December 1916, the Aero Club eventually set up 350 civilian committees, covering virtually every community in the country. More than 1,000 men served on these committees. Large towns had committees "composed of about three influential, public men"; smaller centres, with fewer than 10,000 inhabitants, had one-man operations. In an era when private recruiting leagues provided most of the young men for the Canadian Army, the Aero Club committees performed many tasks that later would be the responsibility of the armed services themselves.[30] But during the latter half of World War I the Aero Club was described as "the backbone of cadet recruiting."[31]

Before the Aero Club could begin its work, RFC Canada launched its own campaign to increase public awareness of military aviation. RFC Canada wanted, above all, to emphasize that its flying training was free. Until RFC Canada began to offer flying training, students had to pay for their own instruction at a civilian school at a cost of about $400, and then had to take their chances on being accepted as a pilot in England.[32] RFC Canada's publicity campaign was directed at "thirty-odd schools and colleges," but was not well known outside of Toronto, and had little success until the Aero Club became fully involved. The Aero Club's civilian committees used the press to promote RFC Canada, which finally began to attract more public notice.[33] As the RFC Canada campaign gathered momentum, the number of applicants increased until it was "absolutely impossible" for the small RFC Canada headquarters staff to interview or even to correspond with all potential candidates. Therefore, it was agreed that the civilian committees would perform the initial screening of cadets, according to standards laid down by the RFC.[34]

The screening process was designed to identify "suitable" prospects, but what constituted a suitable candidate could vary depending on who conducted the interview. One advertisement for "The All-Seeing Aviator" asked for "Clear-headed, keen young men ... possessing a fair education, and sturdy physique."[35] In the words of the history of the Canadian air force, the RFC's ideal candidate was expected to demonstrate "gentlemanliness, educational attainment, mechanical aptitude, and physical excellence, with a measure of recklessness thrown in." Recklessness was to be found in a candidate who "had ridden horses hard across country," or had "nearly broken his neck motoring or ... playing hockey."[36] An interest in "motoring" as proof of recklessness may seem curious today, but it would have seemed logical at the time, because, like an interest in horseback riding, it was a criterion for identifying social class. Automobiles did not become widespread until mass production in the 1920s, and during the war few people had access to a motor car. Knowing how to drive, therefore, often implied wealth and high social status.[37]

But daring was not the only characteristic required of prospective aviators. RFC Canada also limited its selection pool to men who had completed high school with enough credits to permit them to attend university. In addition, an applicant required "letters of recommendation from two prominent public men" who had known the applicant or his family for at least two years.[38] One's parentage was also a matter for official scrutiny, especially if one was not a "European." As in the RFC and the RNAS,[39] only "Europeans" were to be recruited.[40] RFC Canada's attitude toward "non-Europeans" was straightforward: Its application card asked, "Are you a European?" Those who replied "No" were to be rejected. This happened to a black Jamaican, Harold L. Bell, who applied in Toronto to join the RAF as a mechanic during the last year of the war.[41] Another case, that of William H.F. Duke, a Nova Scotia resident and British subject born in British Guiana, illustrated what could happen when the paperwork was botched. Duke, who lived in New Glasgow, Nova Scotia, registered for military service as required by the Military Service Act in November 1917. On completing his application for the RFC, he indicated that he was not of European descent. He was ordered to report to "headquarters" in Halifax anyway, but on his arrival the recruiting staff took one look at him and said, "It's a mistake." Duke was sent home, and in January he complained to the minister of militia and defence about his treatment. The minister asked for an explanation, but the results of the inquiry are not on file.[42]

The exclusion of non-whites from the flying services had its roots in attitudes prevalent at the time. In part the prejudice against non-whites was linked to Social Darwinism, which postulated that natural selection had made white people a superior race. In the 1880s and 1890s Social Darwinism, ignoring technical and military superiority, was used to explain the success of advanced nations (or races, in the nineteenth-century view) against foreign, usually non-white, adversaries. Imperialism became linked to notions of racial superiority, and as H.C.G. Matthew has remarked, the "langauge of 'race'" was used "across the political spectrum."[43] Given all of this, it was easy for many to believe that some ethnic groups made better aviators than others. In 1913, for instance, C.G. Grey in *The Aeroplane* discussed the alleged differences in flying ability among the "Celts," "Normans," and "Saxons" of the British Isles. The official history of the British air effort in World War I stated that "Courage is found everywhere among English-speaking [i.e. white] peoples." Referring to Britons in Arabia, it said that even the private soldiers had shown by their "initiative and tact, their natural assumption of authority ... that they belonged to an imperial race."[44] The policy of excluding non-Europeans from the Canadian flying services was continued until about 1943.[45]

Aside from skin colour, there was a great variety of opinion, among scientists and the public generally, about what constituted the essential qualities of a proficient aviator.[46] Everyone agreed, however, that World War I pilots needed a certain amount of physical strength, for flying could be an extremely demanding activity. John Brophy, a Canadian who trained with the RFC in 1916, described how continuous physical exertion was required to keep a badly rigged aircraft in level flight. James McCudden, a British ace and Victoria Cross winner, remarked that a pilot had to be "awfully strong" to manoeuvre an FE 2b aircraft successfully in combat.[47] Coupled with strength was the stamina of youth. By the end of the war, twenty-two-year-old pilots felt like old men, and the average age of the flyers in some squadrons was less than twenty-two. Anyone twenty-five or older was considered too old to make a good fighter pilot,[48] but Canadians, no matter what their age, had a reputation as above-average pilots.[49] People attributed Canadian success in the air to rugged individualism bred by the frontier, even though most Canadians in the RFC came from an urban environment, and complained vociferously about the cold weather in England.[50] Denis Winter has ascribed superior Canadian performance in air combat to a selection process that placed less emphasis on class distinctions than the British system did, and to a higher Canadian investment in training.[51]

Even though Canadian flyers served with an imperial force, the cultural and geographical distances between this country and Britain led those involved with Canadian aviation into an increasingly close association with their American counterparts. As a result, RFC Canada exerted a strong influence on the beginnings of military aviation in the United States. Later, American developments were to influence Canadian research in aircrew selection. The USAAC experience with aircrew selection thus warrants a brief examination.

A warm relationship (which was to continue into World War II) grew up between Canadian and American aviation officials during World War I, and many aviation training procedures were shared or jointly developed by the two nations.[52] The seminal influence of RFC Canada on United States Army aviation is revealed in the memoirs of Lieutenant-Colonel Hiram Bingham, a college professor and United States Army reservist who helped develop his country's system for the selection and training of military aviators. In May 1917 (the month following the American entry into the war) Bingham visited Toronto accompanied by representatives of six major American universities, and they were shown every aspect of the RFC Canada program. The visitors toured many installations and were given copies of textbooks, of lecture materials, and even of service regulations. Ten days after the Toronto visit the six American universities opened facilities for aircrew selection and ground-school training on their campuses and in so doing they adopted many of the practices observed in Toronto.[53]

With the rapid expansion of its air service, the United States encountered many of the same problems its allies had experienced earlier in the war. As late as April 1917 there were only fifty-two trained flyers in the Aviation Section of the United States Army

Signal Corps (the precursor of the USAAC), but by war's end there were over 16,000 flyers in the USAAC. The pressures of examining thousands of military aviation applicants were handled by the United States in much the same way as Canada and Britain had handled theirs. Although the Americans laid down "rigorous physical and mental standards," including exacting "mental, moral, and professional requirements," the precise means of identifying these characteristics "were necessarily vague and general." Examiners were "instructed to select men of good education and high character," and "constantly enjoined to remember that the flying officer was not to be an 'aerial chauffeur,' but a 'twentieth century cavalry officer mounted on Pegasus.'"[54] But as with the RFC, practice inevitably fell short of the ideal.

In an attempt to solve the examination problem, one of the USAAC's first medical officers, Isaac Jones, an ear specialist, toured the United States to establish recruiting centres and to instruct medical examiners on how to conduct aviation medicals. His ultimate aim was to create standardized tests, so that applicants who failed one examination could not be accepted into the USAAC somewhere else. The USAAC eventually set up sixty-seven recruiting stations staffed by a total of 500 medical experts, and it employed 1,600 doctors by the end of the war.[55] The prominence in American aviator selection of Isaac Jones and other doctors who specialized in physical medicine led to the establishment of standards that focused almost exclusively on the physical well-being of the candidates and virtually ignored their psychological state.[56]

Even though British physicians had been involved in the examination and treatment of combat flyers several years before the Americans, their methods were not significantly more advanced. Throughout most of World War I, medical support for British aviators was furnished by the Royal Navy and the British Army. No specialized medical services were provided to the RFC or the RNAS, and progress in aviation medicine, including selection procedures, depended until late in the war more on the interest of individuals than on any organized effort. Aviation medicine was not recognized as a specialty by the British armed services, and aircrew selection was conducted under the existing medical standards of the military, with only "slight modifications," a situation that went largely unremedied in Britain until February 1918, when formal aeromedical training for MOs attached to the RFC/RAF was instituted.[57]

As the RAF was organized, aviation medical resources were gradually amalgamated, and in March 1918 an administrative committee was formed to control the RAF's medical affairs. The committee was composed of army and naval officers who had been working in aviation medicine, and it was charged with selecting experienced MOs and offering them temporary commissions in the RAF. One of the priorities of the new committee was to scrutinize the work of the medical boards that examined recruits. By the summer of 1918 most of the administrative apparatus of the RAF Medical Branch down to unit level was in place.[58]

Captain C.B. Parker, a CAMC officer seconded to RFC Canada,[59] was one of the first air force physicians from Canada to visit RAF laboratories and medical boards throughout the United Kingdom. Parker was sent overseas to observe British practices with a view to incorporating them into a postwar Canadian aviation medical service. In a letter written in December 1918 he described how he had accumulated "bits of knowledge vis à vis RAF Medical Work" under the guidance of Lieutenant-Colonel Martin Flack, chairman of the RAF's Medical Research Committee. Parker prepared reports on various subjects: the treatment of neurological cases, physical fitness in aviators, the Medical Invaliding Board, "the instruments and appliances of the Aviation Candidate Medical Board," and so on.[60] Both Brefney O'Reilly, RAF Canada's SMO, and another officer – "one of our best men," according to O'Reilly – were scheduled to go overseas to continue Parker's work, but apparently they were unable to do so.[61]

As a result of Parker's reports, Canada's Assistant Director General of Medical Services, Major-General J.T. Fotheringham, suggested to the Chief of the General Staff (CGS) in January 1919 that a small nucleus of Canadian MOs who had specialized in aviation medicine during the war be retained by the CAF. The CGS "strongly recommended" this course of action to the Adjutant-General,[62] but unfortunately for the recently formed Canadian air force, the government favoured financial retrenchment, and the permanent CAF medical staff was reduced to just one officer. Soon even he disappeared, and from March 1923 until early in World War II, the Royal Canadian Army Medical Corps (RCAMC) was responsible for the medical needs of the CAF and the RCAF.[63] Between the wars the small numbers of MOs involved in air force duties, and the transitory nature of their appointments, precluded meaningful Canadian research into the military aspects of aviation medicine, including aircrew selection. This neglect was to have far-reaching effects when Canada went to war in 1939 and needed to raise a large air force quickly.

The postwar reduction of those in the air force performing military duties to a small training cadre spelled the end of the recruiting and selection organization developed in 1914–18. The peacetime approach consisted of offering flying training to small numbers of university students in engineering and applied science, and to a few serving non-commissioned officers (NCOs).[64] Because the small numbers of CAF/RCAF pilots could not justify the cost of maintaining a separate air force selection system, the only formal selection mechanism was the physical examination, based on World War I RFC/RAF standards.[65]

Soon after Canada entered World War II in September 1939, the RCAF began a major recruiting effort aimed at expanding the force from just over 4,000 to more than 16,000 officers and men; in February 1940 the target was increased to 44,500. But the demand for aviators continued unabated and the RCAF continued to grow, its strength peaking in 1944 at 215,200 all ranks.[66] The publicity generated by the exploits of Canadian flyers in World War I helped to ensure that there was no shortage of volunteers to fill RCAF quotas, but expansion early in the war was hampered by the rudimentary state of the air force's recruiting and selection system.[67]

At first, the RCAF recruiting system was overwhelmed by the enormous number of volunteers who came forward; however, measures were soon taken to bring the situation under control. The first step entailed the establishment of recruiting centres across the country. These were staffed by uniformed personnel, so that the first point of contact for new applicants was an air force person, not a civilian, as had often been the case in World War I. The RCAF was able to set up its recruiting centres quickly because of the availability of large numbers of experienced aircrew from the earlier war, some of whom were re-engaged and used immediately to interview new candidates. Although the air force could now process candidates quickly, necessity dictated that selection methods were to remain rudimentary for some time. The first part of the process was an interview, usually conducted by a World War I aviator, who rapidly decided whether the candidate had the required education and motivation. Intelligence was deemed to be "the essential characteristic" of good aircrew material, and education was considered the best indicator of a candidate's intelligence.[68] Education was relatively easy to assess, but many of the other required qualities were less tangible, and for this reason recruiting centres were provided, up to December 1941, with copies of the RAF's Selection Board Notes. These guidelines suggested that the board members should ask questions designed "to throw light on the candidate's breeding and general background," especially if he was likely to be commissioned. "Guts and determination" were to be assessed by "the type of games played."[69]

In addition to the interview, each candidate had to undergo a physical examination, but the shortage of trained doctors created yet another impediment to air force expansion. The situation with respect to medical examinations initially resembled that of World War I; the standards were almost identical to the ones developed in 1914–18 (except that there was no equilibrium test), and the shortage of trained physicians recalled the circumstances of the early part of World War I. Part-time MOs were used at all recruiting centres, where they found it "increasingly difficult" to handle the volume of work. There was a desperate need for doctors trained in aviation medicine to assist in aircrew selection and, from November 1940, to staff the many new RCAF stations established for the BCATP. The military responded by opening the RCAF School of Aviation Medicine in January 1940, and it ran its first month-long qualifying course for MOs in the spring of that year. But it could not produce enough new graduates to immediately fill the many medical vacancies in the RCAF, as every MO already attached to the air force was expected to take the course, which limited the number of places available to others. During the first year of the war the RCAF's medical require-ments were met by doctors seconded from the army,[70] but with the creation, in September 1940, of the RCAF Medical Branch, the air force finally took control of an essential service.[71]

The RCAF Medical Branch initiated measures to alleviate the short-age of qualified MOs, such as increasing the number of courses at the School of Aviation Medicine and waiving formal courses for some physicians,[72] but there were other reasons why the selection process was inadequate early in the war. The flood of volunteers that overwhelmed the RCAF recruiting establishment also placed great strains on the air force's rudimentary administrative system. At Air Force Headquarters (AFHQ), the officer responsible for administering personnel matters was known as the Air Member for Personnel (AMP). Air Vice-Marshal Harold Edwards was AMP from February 1940 until November 1941, and he was succeeded by Air Vice-Marshal J.A. Sully, who held the post until April 1945.[73] The AMP looked after many important areas, such as recruiting, manning policies, pay, promotions, and the provision of medical services. Before September 1939 the AMP branch was tiny, but to cope with the large influx of applicants and to coordinate RCAF mobilization it expanded dramatically in the early months of the war. A new direc-torate, Air Force Manning (AFM), was created on 15 September 1939 to oversee all enlistment and selection activities. AFM undertook many of the tasks performed by the Aero Club in World War I, such as recruiting and publicity. Frequent organizational changes in the RCAF, however, forced Edwards to be continually restructuring his branch, and as late as December 1941 major reorganization was still going on. This complicated the many difficulties that Edwards already faced, one of the most pressing of which was how to best select aircrew trainees from among the large number of aspirants.[74]

Before the outbreak of the war, RCAMC officers seconded to the RCAF had laid the foundations for confronting the selection question. In late 1938 and early 1939 they had approached General A.G.L. McNaughton, President of the National Research Council, and Sir Frederick Banting, Chairman of the Associate Committee on Medical Research, to solicit their support in establishing a committee to coordinate research into aviation medicine on a national basis. With the encouragement of McNaughton and Banting, the Associate Committee on Aviation Medical Research was formed in June 1939, and after reviewing the most pressing scientific problems in aviation medicine, it identified "psychological investigations ... and their application in the selection for flying" as one of the three subjects most urgently requiring attention.[75]

To facilitate its research into aviation psychology, the RCAF soon enlisted the aid of civilian scientists. E.A. Bott, head of the psychology department at the University of Toronto, was the first scientist chosen to supervise this research,[76] which was coordinated by the Subcommittee on Psychology, formed in June 1940, which afterwards became the Subcommittee on Personnel Selection.[77] Bott was "the acknowledged dean of Canadian Psychology." He began teaching psychology in the University of Toronto's philosophy department in 1918 and helped create a separate psychology department in 1926. He was active in many areas of human behaviour, and was a founder and the first president of the Canadian Psychological Association (CPA).[78] Bott regarded himself as an experimental psychologist; he was always tinkering with equipment, and he was "crazy about gadgets."[79] He approached all his research problems by defining their background and context in detail before deciding on a method of investigation.[80] During the 1930s Bott explored the use of "coordination tests" for pilot selection. He conducted trials on Mashburn's Serial Reaction apparatus and McFarland's two-hand coordination device, both of which were designed to measure eye-hand coordination in aviation candidates. Apparently Bott considered the original Mashburn device "completely useless" for pilot selection, and his graduate students derisively labelled it "Bott's complex masturbator."[81] He eventually altered the device and pronounced the modified version "preferable" to the McFarland apparatus for sensory-motor testing.[82] These experiments were to stand Bott in good stead when he was called on to supervise the RCAF aircrew-selection project.

The decision by those who governed and funded aviation medical research to place aircrew-selection research in the hands of a civilian psychologist marked a radical departure from previous practice in Canada. The choice of a civilian psychologist was a result of intense lobbying by the CPA, and prepared the way for changes to the selection process that would incorporate methodologies quite different from those employed by the medical profession. However, there was some doubt that Canadian psychologists were up to this task. Before World War II, psychology in Canada was a young discipline. Still taught as philosophy in some universities, psychology was in the early stages of development in this country. No formal Canadian body existed to regulate the profession, and there was not even a statement of what constituted a practising psychologist. With the clouds of war gathering on the horizon, people teaching psychology at Canadian universities declared their independence from the American Psychological Association in 1938, and banded together to found the CPA.[83] The focus of early CPA efforts was on methods of classifying military personnel, an area of immediate interest to democracies about to expand their armed forces dramatically.[84] The impending conflict was seen by the CPA as a golden opportunity to advance the cause of psychology by offering scientific solutions to some of Canada's military dilemmas.

By the spring of 1939 Bott had formulated a plan for research in aircrew selection.[85] Drawing on his academic background and on his experiments with the Mashburn and McFarland devices, Bott began a search for tests that would meet specified criteria. To be accepted, a test needed to have a scoring system that would be sufficiently sensitive to distinctions among candidates, or in Bott's words that would provide "a wide distribution of scores on a quantitative scale." Bott also had to be sure that a high score on a test would mean "potentially high achievement on the job and vice versa," and that a test would produce "critical score values ... above or below which the chance of success or failure can be defined." Bott wanted to validate tests that would fulfil a number of functions. Some tests were designed to place aircrew candidates into the job – pilot, observer, or wireless operator (air gunner) – that suited them best. Others were designed to predict who would fail pilot training, and still others to assign candidates who failed training for one aircrew trade to an alternative aircrew trade.[86]

Starting in September 1939 and continuing into mid-1940, "about 5,000" candidates were exposed to various psychological methods of evaluating aircrew potential.[87] Selection instruments were chosen for their utility (as described in the scientific literature) and perhaps more importantly, for their availability. Information on a variety of American tests was readily accessible, as Bott had established a close working relationship with the American National Research Council.[88] Most of the methods Bott investigated were based on prevailing theories concerning the evaluation of individual ability, and his first experiments in selection used written tests administered in a group setting and designed to assess "mental ability"; devices to measure manual steadiness and stereoscopic vision; and individual interviews. In 1940–41 tests of aviation aptitude were added. Other attributes were appraised by gathering biographical data from each candidate. Information on a candidate's personal history and aspirations was entered on a special form, and used to gauge his record of adjustment in school and to family life, as well as his emotional and physical adaptability. Then, by means of a structured interview, the candidate's attitudes toward himself and society were probed to weed out the egocentric, and to select those who showed "maturity in accepting responsibility." The interviewer also assessed the candidate's motivation to serve as a member of an aircrew, and his dependability under stress.[89] Finally, psychophysical tests were administered to evaluate reflexes and coordination.

Even though many of the tests showed promise, Bott insisted that they all had to be validated before being adopted for general use. In January 1942, accordingly, "hundreds" of test scores were checked against the results recruits had achieved in aircrew training. The key to Bott's methodology was to allow virtually all candidates who had taken the tests to move on to aircrew training. It was then established that candidates who had attained high scores on certain tests had also done well in training, and that those with low scores on the same tests had done poorly. In this manner, the investigations determined which objective tests were most effective in predicting success in training, and they gradually replaced data from subjective interviews and the candidate's educational records as the basis for aircrew selection.[90]

While working for the RCAF, Bott travelled widely in Britain and the United States for the purpose of examining their aircrew-selection systems. In 1940 and 1941 he produced several reports that detailed the selection methods used in those countries. Bott also recommended components of the tests that could be adopted by the RCAF. His reports were widely distributed, and they appear to have been the reason why he came to the attention of the British Air Ministry at a critical moment for the RAF.[91] The RAF had not made as much progress as its Canadian counterpart in the development of aircrew-selection tests early in the war, for two reasons: a lack of training records in a form appropriate for research purposes, and personality clashes between senior air officers and some leading British psychologists working for the RAF.[92] Consequently, wastage rates in RAF pilot training were unacceptably high. Because all British psychologists had, by mid-1941, been appropriated by the army and the navy, Bott and a University of Toronto colleague, C.R. Myers, were invited to Britain to assist the RAF. They began work for the Air Ministry in December 1941, and, in this unusual way, two Canadians became responsible for directing most RAF research into aircrew training and selection during World War II.[93]

However, their departure for Britain had serious consequences for Canadian aircrew selection. With Bott and Myers overseas, the only senior psychological researcher remaining with the RCAF at the end of 1941 was Sperrin Chant, who was put in charge of the project formerly run by Bott, and given a commission in the RCAF as a Wing Commander. His initial duties consisted of acting as a consultant to the AMP on matters of personnel selection, career development, confidential personal assessments, and the like.[94] Chant was born in St Thomas, Ontario, in 1897. During World War I he served in the Canadian Army, and, owing to his flair for mathematics, he was assigned to keep statistical records for the Director General of Medical Services. He began his university studies after the war, completing his MA in psychology at the University of Toronto in 1924. He also studied psychology at the University of Chicago, where he was influenced by the emphasis on statistical methodologies. On his return to Toronto, Chant joined the psychology department at the University of Toronto and earned a reputation during the 1930s as the department's most accomplished statistician. Chant's forte was the application of quantitative methods to psychology, and he taught courses in statistics and the design of experiments, which replaced earlier courses in philosophy, logic, and the history of ideas. His research included work on factor analysis, the selection of taxi drivers and "men of ability," and the creation of rating scales for clerical staff.[95]

At AFHQ, Chant soon recognized that there was a need for a separate directorate to manage personnel selection and coordinate research and testing in that area. With the support of the AMP, Chant was able to establish a Directorate of Personnel Selection and Research in September 1942, despite restrictions on the creation of new positions at AFHQ.[96]

Chant emphasized that practical results must always be the goal of RCAF research into personnel selection. One of his first assignments was to establish an interim standard for judging success in aircrew training, pending completion of the RCAF's evaluation of psychological testing. A colleague recalled that Chant's "arbitrary" and "rather brutal" – but eminently practical – solution was to "make the criterion whether or not they [the aircrew] actually survive – not figuratively, [but] literally, survive the training period." Chant wanted to "separate those that get killed from those who survive."[97]

While not all BCATP aircrew might have appreciated the design of this particular experiment, Chant was universally admired by those who worked with him. D.C. Williams, a subordinate, commented that "Chant's successful leadership stemmed from his insight into [selection] problems, his administrative skill, and his ability to command respect and establish friendly relations with his fellow senior officers."[98] Chant's contribution to the war effort was officially recognized immediately after the war when, as one of only two Canadian psychologists to be so honoured, he was awarded the Order of the British Empire.[99]

As well as adopting the innovative selection-testing methods developed by Chant and other RCAF psychologists, the Canadian air force also pioneered the use of one of the first flight simulators, the Link Trainer, as a way of assessing large numbers of aircrew candidates. The RCAF's acceptance of the Link Trainer was attributable to initiative and persistence on the part of some air force officers, and to the support of air force psychologists. The Link Trainer was developed in the United States during the 1930s as a device to train pilots in instrument flying, and for its time it was a very sophisticated apparatus. It was mounted on a pedestal and driven by a hydraulic system that could simulate pitch, yaw, and roll. The Link trainer was first used for aircrew selection in 1938–39, when an instrument version was turned into a Visual Link Trainer (VLT) by Flight Lieutenant N.R. Harben of the RAF's No. 16 Elementary Flying Training School (EFTS). Harben removed the blind-flying hood from the standard "D" model Link Trainer and added a painted circular panorama designed to simulate the view from a cockpit. Students were then able to control the VLT with reference to a visual horizon, instead of having to confront the more complex task of maintaining attitude control by instruments alone. Harben's original purpose was to teach the principles of flight to students before they went aloft; however, by identifying candidates with little aptitude for flying at an early stage of their training, the VLT also could facilitate the selection process.[100]

The instrument Link Trainer was introduced into Canada in 1937, when the RCAF installed one at its base in Trenton, Ontario. By early 1938 Link Trainers had been set up at three locations across the country. Technical problems, and doubts about their effectiveness as training devices, curtailed their use before the war, however.[101] At the beginning of the war the RCAF had originally planned to use the new VLTs in the same manner as the British air force, that is, mainly as pre-flight trainers, with Link instructors posted to each Initial Training School (ITS) only to "assist in selecting pilots, air gunners and air observers." However, in October 1940, twelve weeks after the installation of a VLT at No. 1 ITS, Toronto, the Link instructors, on their own initiative, abandoned the idea of using the apparatus solely as a pre-flight trainer and concentrated on improving its utility in pilot selection. From that time on, No. 1 ITS remained the leader of the VLT program in Canada.[102]

From the moment the Link Trainer was installed at Toronto, experiments were conducted to devise a standard pilot-selection test for use with the apparatus; the result was the "Link Test of Flying Aptitude." It required RCAF aircrew candidates (at first during the initial selection process but beginning in April 1943 at the Manning Depots) to complete eight exercises in just over one hour. The testing officer instructed the candidate on how to perform various simulated flying manoeuvres, which the candidate was then allowed to practise twice. This was followed by four recorded trials, in which the candidate's errors were indicated on dials. The dial measurements were added up,[103] and the score that resulted was used to determine whether the candidate was considered suitable for pilot training. Borderline candidates needed to display very strong performance on all other tests and in interviews to justify selection for flying training.[104]

The VLT first came to the official notice of the Canadian psychological community during Bott's visit to Britain in mid-1941. Bott obtained an RAF report that disparaged the device as a predictor of success in the EFTS course, but, based on his experience with psychophysical apparatus in pilot selection, he decided that the criticisms were "sufficiently debatable from a scientific angle" to warrant further study of the subject in Canada.[105]

It appears that Chant became aware of the VLT's potential in late 1941, when he read a report on its use prepared by the officer in charge of the VLT at No. 1 ITS. After several studies under Chant's direction, the most important conducted by D.C. Williams, the VLT was authorized, in August 1942, for general use in RCAF pilot selection.[106] Further experiments by Chant's staff determined that a combination of VLT testing and the RAF's "flight grading" selection method was the most effective way to identify potential pilots. Flight grading simply meant that the student's flying ability was evaluated by a qualified instructor over the course of the first ten or so hours of flying training. The student's performance, and the demand for pilots, determined whether he would continue his training. The RCAF adopted the combination of VLT testing and flight grading as its standard for pilot selection in April 1944.[107] British and American researchers studied the Canadian VLT trials carefully, but they decided not to employ the device for large-scale screening of candidates.[108] Consequently, Canada finished the war as the only Allied nation to employ the VLT comprehensively in pilot selection.

Despite the element of objectivity it introduced into the selection process, the unpopularity of the VLT with both students and staff made its adoption by the RCAF controversial. One reason for its unpopularity was the lack of uniformity in testing methods among selection units. Several steps were taken to address this problem, including the creation of a centralized standards section to ensure that the VLT was used in the same manner everywhere.[109] AFHQ complained in April 1943, however, that "'the Visual Link standardization program is only 50% effective after being authorized for seven months.'"[110] Moreover, the complexity of the machinery demanded highly skilled maintenance personnel to keep it running. Technical problems and a dearth of technicians contributed to the VLT's difficulties, which continued to dog the apparatus even after the end of World War II.[111] Nevertheless, refinements in the equipment and in test procedures gradually improved the VLT's ability to predict performance in pilot training. Preliminary testing of 1,000 students at No. 1 ITS in 1941–42 was found to have "accurately predicted 92.5% of the successful pilots and 48.5% of the failures."[112] Another study, published in 1943, found VLT scores to be very accurate in predicting success in flying training up to and including the Service Flying Training School (SFTS) level. Drawing on VLT scores at No. 1 ITS, the study found that Link scores had a very high correlation (coefficient of 0.57) with success in pilot training.[113] A 1944 study of 366 aircrew reported a correlation coefficient of 0.41, which placed VLT testing eighth among twenty-seven "psychomotor and information processing tests" reviewed by aviation psychologists in the 1980s.[114] The strength of the 1944 investigation, which was known as the "Arnprior experiment," lay in the fact that it was carefully conducted under the supervision of Flying Officer Erdo Signori, a trained AFHQ psychologist. The Arnprior experiment was designed so that even candidates who scored low on the VLT were given flying training, to judge whether there was a correlation not only between high scores and success, but between low scores and failure. Signori concluded that the RCAF's Link-based system combined with the RAF's flight-grading method was the most effective way to identify pilot aptitude. The Arnprior experiment is still cited today as an indicator of the usefulness of this hybrid pilot-selection procedure.[115] Whatever the VLT's weaknesses were, it proved its worth in pilot assessment, and thus contributed to reducing wastage in the RCAF.[116]

While delays in implementing a standardized VLT program throughout the RCAF meant that most BCATP applicants passed through aircrew selection before research into test procedures could be completed, it could be argued that, once psychologists became involved, the selection system was developed as quickly as was practicable. However, it did take several years to completely implement procedures that correlated well with success in training. This can be explained by the fact that a time-consuming controlled-research program, involving thousands of subjects, was necessary to produce the procedures.

By war's end, Chant and his colleagues had developed an aircrew-selection system that compared favourably with others in use around the world.[117] Chant asserted that "disproving the validity of certain cherished procedures and forthwith discarding them has been one of the most valuable features of the Air Force selection programme."[118] By this he meant that he had used scientific methods to establish that there were more effective selection procedures than the subjective assessments preferred by some RCAF officers. By virtue of his privileged position at headquarters, Chant was able to build on Bott's research into objective testing methods and to insist that selection criteria based on individual prejudices be replaced by procedures validated through experimentation.

But despite its successes, the RCAF's aircrew-selection program still had problems. A technocratic élite – psychologists – came to dominate the program, and their motives for participating in it were self-serving as well as patriotic. No doubt Canadian psychologists were guilty of some of the sins enumerated by Stephen Jay Gould, including the reification of complex human capabilities, the use of relatively simple criteria to rank people, and allowing unconscious biases to affect research.[119] But they did anticipate some of Gould's objections to the "scientific" measurement of human potential. For example, they accepted the premise that intelligence was not an immutable quality and that it could be augmented by education. In fact, whereas the United States Army psychologists of World War I, according to Gould, used the results of intelligence tests to justify a hereditarian ideology, Canadian psychologists determined that there was a weak correlation between the RCAF Mental Ability Test and success in pilot training.[120] Chant and his colleagues would probably have agreed with Gould's caution against trying to gauge the complexities of human behaviour numerically. But with a war on some short cuts were required. Chant spoke for many psychologists working in the area of aircrew selection when he enunciated, in 1943, what he called "one of the basic principles" of psychological assessment, namely that "totalities are not amenable to measurement." The system he developed was thus founded on tests that "most closely" simulated "the actual trade for which the individual [was] being selected."[121] By not reifying certain general factors, such as intelligence, and relying instead on criteria that approximated as far as possible the real business of flying, RCAF psychologists avoided some of the worst of Gould's pitfalls.

But psychologists were not the only professional group to face the challenges presented by aircrew selection. The medical community, as we have seen, contributed to aircrew selection in World War I, and it continued to play an important role in World War II. From 1941 to 1944 the RCAF rejected only 22 per cent of aircrew candidates for physical reasons, which was one-half the army's rate. The official history of the Canadian medical services ascribed the smaller proportion of rejections to the voluntary nature of RCAF enlistment, and stated that it was "in no way indicative of ... [the health of] a cross-section of young Canadian males of the same ages.[122] The official history assumed that many of the country's healthiest and best-educated young men were present among the large numbers of volunteers for aircrew duties. But it must also be said that physicians had become more effective since the early days of aviator selection in sorting out who had the necessary physical attributes for service in the air. Nevertheless, a number of medical tests remained in place during the first two years of the war that were responsible for rejecting significant numbers of capable candidates. Canada's abandonment of its own air force medical branch after World War I left it dependent on RAF medical criteria for aircrew selection for most of the next war. For instance, even though the standard aviation medicine textbook of the years between the wars rejected the World War I selection procedure that required candidates to support a column of mercury by blowing into a tube, that procedure was still retained by the RCAF until at least early 1942.[123] Together with the unduly stringent eyesight requirements imposed by the RCAF – suitable only for tiny interwar air forces able to tolerate high rejection rates – it was responsible for the loss to the air force of much valuable manpower. From August 1941 to June 1944, for example, "visual defects" alone disqualified 51.8 per cent of all RCAF aircrew candidates.[124]

However, those responsible for Canadian aircrew selection in World War II eventually overcame many of these problems and transformed the process from one based primarily on a medical model to one that incorporated broader tests of aptitude. Physical fitness testing was the initial screen through which successful applicants had to pass, but psychological tests were used to assess their chance of completing training and to determine the aircrew category into which they would be accepted. Using Bott's prototypes, the RCAF also developed standard tests designed to measure intellectual capacity and learning ability, as a means of evaluating candidates with limited formal education. But arguably, the greatest wartime achievement of the Canadian psychologists was the integration of the VLT – a reliable predictor of success in pilot training – into the RCAF aircrew-selection process.

From the point of view of selection philosophy, perhaps the most noteworthy change from World War I methods was the use of statistics to establish a sliding scale of assessment criteria in place of discrete pass/fail criteria.[125] The test administrators of World War I unofficially recognized the need for a sliding scale by sometimes altering the grade required for a pass, and, more often, when the demand for aircrew was high, by allowing failed candidates to retake tests. It was only in World War II, however, that statistical methods were used to rank candidates from best to worst, and policies devised to choose candidates of the desired aptitude level to meet the demand for aircrew.[126]

Despite the selection system's flexibility and its success in reducing the numbers of those who ceased training, its effectiveness was open to challenge because it did not measure the ability of aircrew candidates to endure the stresses of combat. The problem of selecting aircrew with the right fighting temperament – an issue first identified in World War I – was never really addressed by the RCAF in World War II. During World War I, attempts to correlate operational success with easily identifiable selection criteria had been confounded by the difficulty of defining what really constituted operational success.[127] Although some work was done late in World War II to establish criteria for predicting operational success, researchers were hindered by the same conceptual problems that had plagued their counterparts of 1914–18.[128] Some improvement in operational loss rates during World War I had been achieved, however, by using higher physical standards to eliminate people who were clearly unable to make effective combat flyers. In 1915, 90 per cent of RFC losses were attributed to pilot performance, and 60 per cent of the pilots in this group had a physical defect that significantly affected their flying ability. The adoption of strict medical standards reduced RFC casualties linked to physical unsuitability for flying to 20 per cent in 1916 and to 12 per cent in 1917.[129]

The World War I aircrew-selection system has been characterized as one that focused on "general merit," picking out those men who would do fairly well at almost anything.[130] In World War II more specialized selection tests were designed to specifically predict success in training. If World War I procedures were designed to favour generalists, the procedures of the next war separated those who had the best chance of completing training from those with good promise overall.

This chapter's account of Canadian aircrew selection from 1914 to 1945 has illustrated one case of a "managerial metamorphosis" in warfare, where the old ways of doing things were swept away by technocratic élites. The ad hoc selection of aircrew trainees by well-meaning amateurs at the start of World War I was supplanted in part by the centralized examination system adopted by RFC Canada in 1917, and in World War II the entire selection process was run from the beginning by the air force. Medical doctors constituted the first technocratic élite to oversee aircrew selection, but in Canada after 1940 they lost out to the psychologists, chiefly because it was found that physiological assessments alone did not give as complete or as accurate an indication of aircrew potential as did a combined physiological and psychological evaluation. The ability of civilian psychologists such as Bott and Chant to achieve practical results through the use of new statistical techniques gave these specialists power in a sphere that had been the preserve of other experts. Moreover, the development of distinctive Canadian aircrew-selection procedures established national autonomy in a vital area of man-power conservation and produced a system that endures fifty years later.

3 Lessons Learned, Lessons Forgotten: Aircrew Training, 1914-45

Success in air warfare depends heavily on the skills of those who operate the aircraft. The training of flying personnel is influenced by many factors, including political considerations, aircraft availability, human-resource and selection policies, strategic and operational requirements and wastage. In war, all of these factors interact in a complex and constantly changing manner to determine the quality of the final product – the trained flyer.

One of the greatest difficulties in designing a system of flying instruction in World War II was that one to three years were needed to train aircrew to operational standards, during which time it was impossible to predict how changes in the course of the war might affect instructional requirements.[1] The complications are evident when one considers that decisions made during 1942 – the nadir of Allied fortunes – committed air resources to a course that was to last until 1944 and have results that could not be clearly foreseen.[2] The decision of the British Chiefs of Staff in October 1942 to accept the heavy bomber as the principal Allied weapon against Germany,[3] for example, was to have far-reaching consequences for the BCATP. Once underway, the training juggernaut developed a momentum of its own, and, while its course could be deflected to a small degree, it could not be radically changed.

Flying training, like most kinds of learning, must strike a balance between quality and quantity. It requires time to produce high-quality flyers, and if the training period is reduced to meet a demand for more aircrew, the result is invariably a higher wastage rate, both in training and in operations. At the same time, the demands of the training system can have an adverse impact on combat effectiveness. For example, of the 5,300 medium and heavy aircraft in Bomber Command in early 1943, more were being used for training than were in action against the enemy in occupied Europe.[4] To maintain both his command's expansion objectives and its sortie rate against the enemy, Arthur Harris was forced to withdraw "88 per cent of his pilots to feed OTUs [Operational Training Units] and Conversion Units before completion of their first tour."[5] To add to this burden, as the war progressed technological advances increased the complexity of air warfare, and the need to impart more skills and knowledge to Bomber Command aviators had to be weighed against the effects on operations.[6] More time could have been allotted to increasing crews' proficiency with new equipment, but this would have slowed the flow of replacements to, and thereby weakened the combat capabilities of, the front-line units, which were suffering heavy casualties.

A fair amount has been written about the history of aircrew training,[7] but the following discussion approaches the subject from a new perspective, by looking at how the evolution of aircrew-training practices affected air force manpower policies, especially in Canada, from 1914 to 1945.

In the earliest days of aviation, flying lessons were given by private individuals who ran their own schools equipped with a handful of aircraft, often of their own design and construction. No organizations existed to regulate standards of instruction, and the tutelage could be extremely unsophisticated. Lanoe Hawker, who eventually became one of the highest-scoring pilots in the RFC and won a Victoria Cross, learned to fly at the Deperdussin school at Hendon in the spring of 1912. Hawker called the Deperdussin teaching methods "tedious." Students began by taxiing their aircraft about, gradually increasing speed until the wheels just left the ground. This process was repeated, with each "hop" being a little longer until the student got up the flying speed, and the nerve, to do a circuit of the aerodrome. Progress was slow as few aircraft were available, and novices were only allowed to practice in a "dead calm." Because little was known about the principles of flight, and aircraft with dual controls were not available at Hendon, the actual amount of teaching provided was minimal. As well, each instructor "had his own ideas on flying, and it was often a matter of chance if the pupil did the right thing in an emergency."[8] The Hendon experience was typical of many schools, but not all the instructors at Hendon were quite so nonchalant as Hawker's. L.A. Strange, who learned to fly at about the same time as Hawker, had a very meticulous instructor who insisted that his students practise simulated flying manoeuvres on the ground (in an aircraft mounted on trestles) before getting airborne.[9]

But no matter how they were taught, all students had one aim: to acquire the licence granted by the Fédération aéronautique internationale (FAI) through its national branches. The test for the licence consisted of taking off, climbing to about five hundred feet, completing two successive figures-of-eight, landing, again taking off and climbing to about five hundred feet, making an approach with the engine off,[10] and touching down in an area about the size of a tennis court. In Britain, the test was conducted by an official of the Royal Aero Club, who, rather than joining the student aloft as a passenger, stood on the ground and watched the proceedings from a safe distance – a wise precaution given the minimal flying experience of some students. In 1914–15, most students took the test after just four or so hours of flying time, roughly one hour of which was solo. The record for shortest time in the air before taking the test appears to have been earned by a student who passed his test with "just under ten minutes' flying time" in his logbook.[11]

Until August 1916 the FAI licence was the only flying qualification that prospective military pilots accepted by the RFC were required to have.[12] The RFC assumed that anyone who held the licence had the necessary skills to immediately begin military flying training. But this was far from the reality of the situation, for many pilots went to the front still learning to fly their aircraft. Because of the lack of experience of its pilots, RFC casualties soon became insupportable, and eventually provoked drastic changes in the training system. At first, little was done except to bring seasoned combat pilots back from the front to share their experiences with new pupils. By the end of 1915 specialized schools had been established to teach aerial gunnery and fighting, and wireless operation. Then, in early 1916, RFC training methods were completely revamped. Before 1916, the RFC had used several methods to train aviators. One was to form training squadrons (known as "reserve" squadrons) to perform two functions at the same time. First, the squadron members literally taught themselves to fly, with the best pupils joining the few experienced flyers as instructors as soon as they were able. Second, new service squadrons were created by hiving off personnel, novice and skilled, from the reserve squadrons to form fresh units destined for France.[13] The haphazard practices of 1914–15 were gradually replaced by a formal training scheme in which the required hours of flying instruction were increased and wings awarded only after tests simulating wartime missions had been passed.[14]

The new training format helped reduce losses at the front, but until instructional techniques were further improved, the flying schools continued to be a source of serious wastage. It is difficult to arrive at precise numbers for losses incurred during flying training in World War I, but the estimate of 28 per cent for late 1917 given by the official British history of the war in the air is a representative figure.[15] Besides the lack of a proper syllabus, the poor quality of the instructors also contributed to training losses. Before steps to improve teaching methods and to select better instructors were instituted in late 1917, it was not uncommon for flying-school staffs to include men whose nerves had been shattered by their experiences in France, and others considered incapable of active service. And in those cases where flying-school graduates were posted immediately to instructional duties, no effort seems to have been made to choose the best-qualified flyers as teachers.[16]

For these and other reasons, there were many accidents during training, although fortunately most of them were not fatal. A typical day at a training camp saw three to seventeen crashes resulting in aircraft damage and the pilot's injury or death.[17] One reason for the many mishaps was the semi-official encouragement of dangerous stunt flying as a way of inculcating "offensive spirit." Vivian Voss, who underwent pilot training at Camp Borden in the summer of 1917, recalled that his commanding officer did not "compel students to stunt," but would have been "disappointed if you d[id] not attempt *some* stunts at Borden."[18] When RFC Canada trained in Texas during the winter of 1917–18, "'the death rate was out of all proportion'" to the hours flown, according to one participant, because "'[t]here were no curbs on any type of flying; every good stunt ever heard of was attempted by anyone crazy enough to try it.'"[19] And not only students and instructors were killed; many experienced combat pilots were also lost to "senseless stunting."[20] The fate of an unnamed British ace with more than fifty victories to his credit was typical. On his first visit to an American training base, he amazed onlookers with breathtaking aerobatics, including loops so close to the airfield that his aircraft's wheels came "within a few inches of the ground." He was killed soon after when he hit a hangar while performing similar manoeuvres. The American base commander commented that British squadrons had "far higher morale" than American squadrons because of stunt flying, but he lamented the needless loss of life.[21]

While it may have been difficult to curtail the antics of front-line pilots, it was possible to influence the behaviour of trainees. What was required was a systematic, standardized method of flying training to impress the notion of air discipline on novice pilots. The

RFC eventually implemented such a program, and it became known as the Gosport method of flying instruction.

Named after the aerodrome, just west of Portsmouth, where the system was perfected, the Gosport method was the brainchild of Major R.R. Smith-Barry,[22] a student of the theory of flight who concluded that flying could be explained by relatively simple physical principles. Before Smith-Barry's procedures were adopted, students simply copied their instructors' actions, with little understanding on the part of either of the principles behind the control movement.[23] The Gosport method emphasized explaining the physical laws of flying manoeuvres to students, demonstrating the manoeuvres in a dual-control machine, and then allowing students to practice the manoeuvres in the air, accompanied by an instructor at first, and later on their own. Smith-Barry's procedures were propagated throughout the RFC/RAF in service publications, and, more importantly, beginning in August 1917, by the establishment of schools for flying instructors.[24] Eventually all instructors had to pass through the schools, where the "slip-shod teach yourself method" was replaced by the new formal systems. Many experienced combat pilots, on their way to becoming instructors after a tour at the front, discovered at schools using Smith-Barry's procedures "how very badly they really fl[ew]."[25] In the Gosport method, Britain had one of the most progressive pilot-training techniques in the world. American authorities, who used it at some of their schools in preference to the French technique (which resembled the pre-war Deperdussin approach), recognized the superiority of the British system.[26]

Word of Smith-Barry's innovations reached Canada in March 1918, but it was not until July of that year that a significant number of Canadian instructors became familiar with his method, which had recently been adopted for use in Canada.[27] The Armour Heights System, named for the airfield on the northern outskirts of Toronto where it was taught, was RFC/RAF Canada's adaptation of the Gosport method.[28] In the Canadian program, each instructor was limited to six students, and held responsible for any crashes, on the theory that it was his job to make competent solo flyers of his students, and, in the early stages of training, when dual-control aircraft were used, to prevent dangerous mistakes.[29] In April 1918 a School of Special Flying was established at Armour Heights for novice instructors; eventually, all instructors were required to take the course.[30]

Perhaps the most important innovation of the Gosport system was the creation of the post of "wing examining officer," who today would be called a standards officer or check pilot. In RFC/RAF Canada, each wing had an examining officer whose duty it was to "weed out those [instructors] who got stale" and to "test the abilities of all new instructors from time to time." He was also charged with investigating student crashes to see whether particular instructors were responsible for more than their fair share.[31] The Armour Heights System appears to have cut the crash rate in half from July 1918 to the following November.[32]

But flying training was simply the final stage of the comprehensive system maintained by RFC/RAF Canada to prepare its aviators for war. New entrants began their career in RFC/RAF Canada at the Recruits' Depot (opened at Toronto in February 1917), where they underwent a final medical inspection and were issued their kit.[33] The earliest recruits received no formal ground-school classes, but proceeded directly to flying instruction. In March 1917 a new training body, called the Cadet Wing, was established at the University of Toronto, and a full ground-school program was taught there. With the addition of trained staff from Britain in the spring and summer of 1917, Schools of Military Aeronautics and Aerial Gunnery, and an Armament School, were set up to provide advanced courses for fledgling flyers.[34]

By the spring of 1918 RFC/RAF Canada had developed a coordinated course of study which was taught in a prescribed sequence at the various schools under its control. After no more than two weeks' basic training at the Recruits' Depot, the student was sent to the Cadet Wing, which had moved to Long Branch, a few miles southwest of Toronto. There he spent eight weeks improving his knowledge of wireless operation, learning how to read signal lights and panels, and being introduced to air navigation. Next, he was exposed to the battlefield applications of the knowledge he had acquired, by learning, for example, the complex principles of artillery cooperation, which were taught at the School of Military Aeronautics, now located at the University of Toronto. Here he also received further general military instruction, and was introduced to the mechanical aspects of aircraft engines and to the scientific principles of flight. The School of Military Aeronautics course was set at six weeks, and no cadet was to continue on to flying training until he had achieved a grade of at least 65 per cent on his written examinations. Next, the student spent four to five weeks at the Armament School at Hamilton, where he learned the theoretical and practical aspects of aerial gunnery and bombing. He then went to one of the three Flying Wings (North Toronto, Camp Borden, or Deseronto), where he got his first opportunity to go aloft. A course at the School of Aerial Gunnery (later the School of Aerial Fighting) at Camp Borden, where the new "science" of air combat was taught, completed his training in Canada.[35]

While the first aircrew-training facilities were primitive, because of the rush to get them into service, one cannot help but be impressed by subsequent improvements. The semi-official history of RFC/RAF Canada by Alan Sullivan contains many photographs of the sophisticated battlefield models, aircraft, and training devices used in the later stages of the war by the Canadian schools, which suggests that the RFC/RAF Canada training system would compare favourably to the best schools in the Canadian Forces today. Some of our less enterprising establishments still have something to learn from their forebears' industry and imagination.

The official Canadian air force history has noted that RFC/RAF Canada had a very strict policy on discipline, but was quite lenient toward cadets who had trouble passing the training course. Cadets were routinely punished and even discharged for misbehaviour, yet they were given many opportunities to retake ground-school and flying exams.[36] This might be attributed to the army attitude that discipline was more important than technical proficiency, an attitude that was, over time in the air arm, to give way to the concept of the airman as a pilot first and an officer second.[37]

The improvements made in Canadian aviator training during World War I significantly reduced wastage; fatal accidents, for example, decreased significantly after the introduction of the Gosport system into Canada. But even at its best, RFC/RAF Canada's fatal accident rate was two to four times greater than the BCATP rate in the next war.[38] This can be attributed to a number of causes, but perhaps the most critical was the greater unreliability of the aircraft of 1914–18, especially their engines.[39] Of almost equal importance was an indifference to the dangers of stunting and equipment damage that is difficult to comprehend today. The hazards of World War I flying training were exacerbated, moreover, by the pressures placed on the training system by the British Air Ministry for the replacement of flyers killed at the front, particularly in 1918 when the casualty rate was running at 32 per cent a month.[40]

By the end of the war, a flourishing aircrew-training system was operating in Canada. At its peak, in 1918, it was producing about one-fifth of the aircrew-reinforcement needs of the British Empire. By the signing of the armistice, more than 3,200 flyers had graduated from the Canadian program, an outstanding achievement for a nation of Canada's population and industrial base. Even more significant, from a Canadian point of view, was that by November 1918 two-thirds of the staff positions and 70 per cent of the flying positions of RAF Canada were filled by Canadians, thereby establishing the first large cadre of flying instructors in the country.[41]

By war's end Canada had the infrastructure and the personnel to successfully conduct aircrew training on a large scale. But like all other aspects of military life in Canada, the training establishment atrophied with the coming of peace. From late 1918 to 1923 the only training offered by the CAF was in the form of refresher courses for those who had earned their wings during the war. As this group began to settle into middle age, however, it was decided to begin recruiting new pilots.[42] The interwar period witnessed the creation of two pilot-training programs. The first, the Provisional Pilot Officer (PPO) scheme, began in May 1923. This plan was the vehicle by which the vast majority of Canadian military pilots were trained between the wars. At first, only university students in science and engineering were offered flying training by the CAF/RCAF. The authorities felt that, because the air force was "in a sense, a very large engineering organization," only such graduates would be qualified to apply for commissions in the Permanent Force. For most of the PPO scheme's history, cadets took their training over the course of three summers. The intermittent nature of the program accounted in large part for its high attrition rate of 58 per cent. Under a new PPO scheme, introduced in October 1932, candidates were appointed temporary PPOs after completing university; they then received eight (later twelve) months of continuous training.[43] This change in structure appears to have been responsible for a reduction of the attrition rate to 31 per cent in 1932–40.[44]

A second program offered pilot training to NCOs. The first course began in February 1927, and by May 1931 the program had graduated thirty airmen. The attrition rate was 33 per cent, almost identical to that of the PPO scheme in 1932–40. It was then suspended for five years, but started up again in the spring of 1936. By the outbreak of World War II, a further twenty-six NCO pilots had been trained.[45]

During the interwar years, flying training in the CAF/RCAF initially appears to have followed the well-tested World War I methods with which the senior instructors, veterans all, would have been familiar.[46] A notable change occurred in 1928 when, in the interests of economy, the RCAF entrusted primary pilot training to fifteen civilian flying clubs across Canada; by 1939, twenty-three were providing this service. The clubs were given two aircraft each and a $100 subsidy for each pilot trained.[47] To assist the clubs, the air force offered flying-instructor courses to civilian pilots,[48] and, beginning in 1933, sent two of its officers across the country to test and certify civilian instructors.[49] As a further economy measure, responsibility for teaching of the advanced-flying syllabus was transferred from the flying schools to squadron-level instructors, a procedure copied from the RAF. In

1934 RCAF expansion put even greater pressure on limited training resources, and additional flying-instruction responsibilities were handed over to the squadrons.[50]

On the eve of World War II it was decided to employ one method of pilot training within the Commonwealth, and in May 1939 the RCAF adopted the British air force syllabus, which divided flying training into three stages, elementary, intermediate, and advanced. Each stage was about sixteen weeks long, and one year was allotted to complete the entire course. In Canada the civilian flying clubs continued to provide elementary instruction, and to encourage their participation the government agreed to supply two aircraft for every one provided by a club, and to pay the clubs $840 for each pilot they trained. To ensure uniform training methods, all civilian instructors were required to take a refresher course at the Flying Instructors' School (FIS) that had been opened at Camp Borden in April 1939.[51]

During the interwar years the RCAF was able to preserve some of what it had learned about individual flying training in World War I. From 1919 to 1939, however, standards were continually being diluted. The decline began with the decentralization of elementary training to the many civilian schools, and continued when advanced training was entrusted to individual squadrons.[52] Another reason for the decline was the absence of a separate standards organization between the wars to closely monitor flying instructors and to review the training process.[53]

Compounding these difficulties was the RCAF's dependence on the imperfect British model of flying training. After 1918 the job of teaching pilots in the RAF "became a settled, regular, familiar routine."[54] According to one study, "CFS [Central Flying School] kept flying instruction at an admirable and polished standard ... but only within narrow limits," and it was almost solely concerned with producing pilots skilled in short-range flying. By the early 1930s the curriculum had "become static and stereotyped" in the mould of the last war. Modifications were introduced in 1934, but considerations of "economy and immediate expediency" interfered with their implementation. Moreover, as one way of employing squadrons usefully in peacetime, and to reduce the costs associated with maintaining a large training establishment, the RAF had gradually devolved much of the tutoring of novice pilots to operational squadrons. As a result, training soon became viewed as a secondary, nonessential activity, and by relegating it to a subordinate place during the interwar years, the RAF significantly diminished its efficacy and importance.[55] Finally, during the second half of the 1930s the RAF was unable to provide sufficient staff for new flying schools without seriously depleting front-line squadrons. This kept instructional development stagnant as war approached, and was a "powerful retarding influence" on the air force's adaptation to technological change.[56]

These impediments to effective training were recognized by Air Commodore (later Marshal of the RAF) Arthur Tedder, who advised in 1934 that any increase in the strength of the RAF should be based not, as had happened in World War I, on an ad hoc expansion scheme, but on a "'reasoned training expansion programme.'" He insisted that a higher level of qualification had to be provided by the flying-training schools, and asserted that unless growth was tied to instructional capabilities, "'we shall again fall between two stools and secure neither the squadrons we want nor the training which is requisite.'"[57]

Unfortunately for the RAF, its failure in 1939 to heed Tedder's warning led it down the same path it had trod in 1914–18. In 1939 the RAF "was relying on a training system which remained unchanged in scope since the early 1920s, except for a patchwork of enforced additions and extensions." This was to have serious repercussions during the first two years of the war.[58] In this critical period the RAF's atrophied instructional organization was asked to produce large numbers of aircrew trained in the new skills necessary to operate modern aircraft, but as air warfare rapidly evolved, the skills taught in the schools could not always keep up with operational requirements. The full measure of the problem was not evident until the summer of 1940, during the invasion of France and the Battle of Britain, when RAF Commands found that economy in operational training had "crippled" the front-line squadrons. In 1941 the RAF initiated a "New Deal" to correct the problem. By increasing the hours of tutelage in the air that students received, it aimed to significantly improve the quality of aircrew trainees. However, the plan only attained the goals proposed by Tedder in 1934.[59]

In addition to its internal deficiencies, the RAF training system was beset by external pressures that made its goals much more difficult to achieve. Many people believe that the Allied victory in World War II was inevitable, but it is well to remember that the war did not clearly turn in the Allies' favour until the latter half of 1943.[60] For two-thirds of the war, therefore, victory was by no means certain. One of the most insistent challenges faced by Britain's aircrew-training establishment was the need to adapt to rapid changes in combat priorities and in Air Ministry policies. After Dunkirk the greatest demand was for fighter pilots, and this remained the case until June 1941, when almost all German bomber forces were withdrawn from operations against Britain and committed to the invasion of Russia. From that point on, priority for aircrew was generally given to strategic bomber forces, which were seen as the war-winning weapon. However, their build up was significantly slowed by the competing demands of the Middle Eastern theatre in 1941–42. In late 1942 aircrew to combat the u-boat menace became the new priority, and after D-Day the need for close air support meant that priorities changed again. The net effect of these changing operational requirements was that the training program was never able to achieve stability. For the RAF (and the RCAF) this caused, in the words of F.J. Hatch, "a gigantic problem," as course syllabi were in a state of continual flux, and it sometimes seemed that no one was in control.[61]

Although changing operational requirements presented formidable challenges to the training system, the chief bottleneck from 1941 on, especially in Bomber Command, was the OTUs, for three reasons. First, by the middle of 1941 the number of pilots trained to the pre-OTU standard was no longer – and never again was really to be – the limiting factor in expansion; in fact, embarrassing surpluses soon appeared. Second, in mid-1942 the introduction of new electronic devices meant that more time had to be spent on specialized training, which could only be provided at the OTU level. Third, Bomber Command OTUs were understandably loath to graduate students until they reached a minimum standard, and this resulted in some crews remaining on course three times longer than the expected six to eight weeks. In late 1941 the Chief of the Air Staff (CAS) reluctantly approved the practice of extending the courses in certain cases, because operational losses due to inadequate training were becoming intolerable. Low serviceability rates (down to 50 per cent in late 1941), the awful weather of the winter of 1941–42, and the poor calibre of instructors compounded the OTUs' problems.[62] Until training improved, first-line squadrons in Bomber Command were forced to spend 40 per cent of their flying time on non-operational exercises to bring new crews up to an acceptable level of proficiency.[63] In April 1942 all of these considerations finally prompted the Air Ministry to order that heavy bombers would carry only one pilot instead of two. The new policy allowed heavy-bomber OTUs to more than double the number of flying hours each student pilot received.[64]

But this had a deleterious effect on the BCATP, which had already geared up to produce the maximum number of pilots to meet earlier forecasts. As a result of the Air Ministry order, the Canadian training program had to be restructured to produce fewer pilots and more aircrew of other kinds. This is just one instance of how British operational decisions affected the RCAF's training role.

Beginning in 1939 the RCAF built a training infrastructure on a scale unprecedented in Canada.[65] An immense construction program and the formation of instructional cadres occupied the full attention of the RCAF's Air Member for Training (AMT), Air Commodore (later Air Marshal and CAS) Robert Leckie. His department expanded dramatically from a small pre-war training group to four training Commands. Because of the Commands' geographical dispersion and the small size of AFHQ, it was decided that "each command was to be as self-sufficient as possible, with its own recruiting and manning organization ... and air training schools."[66] This policy was to have important consequences for the production of aircrew in Canada, as we shall see in a moment.

Generally speaking, the RCAF turned out well-prepared aviators throughout the war; however, there were wide variations in standards over time and between BCATP schools,[67] for a number of reasons. First, as was mentioned above, the war placed continually changing demands on the system. For example, on the outbreak of the war the RCAF was asked to produce 2,000 pilots annually and "as many observers and air gunners as possible," but by April 1942 the number had soared to 19,500 aircrew a year.[68] Secondly, as in 1914–18, the pressure to meet quotas had an adverse effect on the quality of the graduates. Before 1944, when producing large numbers of aircrew was the priority, school commanders were routinely removed for not achieving the numerical goals set by their superiors.[69] A lowering of graduation standards was the inevitable result. Afterwards, when fewer aircrew were required, the quotas were reduced, failure rates were deliberately increased, and only the best students were permitted to complete their BCATP courses. The third reason, which will be discussed in detail later in this chapter, had to do with changes in training standards.

For the typical aircrew candidate, progress through the different stages of training was long and arduous. After satisfactorily completing initial selection procedures at the recruiting centre, typically the RCAF Classification Test (designed to measure learning ability); a medical; and an interview to assess his motivation to become aircrew, the candidate was enlisted as Standard Aircrew, with the rank of aircraftman second class (AC2), and posted to a Manning Depot. At the Depot he was given basic military training and tests of aptitude and educational achievement. If he was found to be below the required educational standards, he might be sent on an academic course for six to twelve weeks to bring him up to the necessary level. A more thorough aircrew medical was also conducted at the Depot. An Aircrew Selection Board then considered his test results and choice of aircrew category, taking into account the air force's requirements at the time. Men chosen to be navigators and air bombers were sent to ITS to begin instruction in ground subjects before moving on to the specialist schools, while prospective wireless operators (air gunners), air gunners, and flight engineers went directly to the specialist schools. After completing ITS, student pilots proceeded to EFTS and SFTS to complete their basic flying training. All candidates who received their wings were promoted to the rank of sergeant, and some were commissioned soon after. Those aircrew who were selected for operational assignments were sent to OTUs or to other advanced training before receiving their squadron postings.[70] With the delays caused by travel and by waiting for vacancies in advanced courses, it was not unusual for aircrew training to take up to two years.

Over one hundred schools, teaching a broad range of subjects and incorporating the latest advances in technology, were established by the RCAF to meet the training needs of the BCATP, and there were many obstacles to be overcome in organizing an educational system of such size and complexity.[71] The small size of the pre-war air force, and the immediate demand for experienced officers in command and administrative roles, caused the RCAF to staff its schools with inexperienced instructors. Ground-school training at the ITSs suffered, until the summer of 1943, from poor instruction and a "lack of uniform standards."[72] The flying-training schools, with a smaller pool of potential instructors to draw on, faced even greater difficulties. They were forced to obtain their staff from among civilian pilots and fresh graduates of the BCATP.[73] These instructors were generally not experienced flyers, some being accepted by the RCAF with as little as forty or fifty hours of flying time.[74] At the EFTSs most of the civilian instructors had only one hundred to one-hundred-and-fifty hours of flying time, and many were simply given the so-called Gosport-patter test before being accepted by the RCAF. Knowing the Gosport patter meant memorizing and reciting by rote mimeographed notes based on a corrupted version of the Gosport method of World War I.[75] The result of all of this was a system that produced inexperienced and often poorly motivated instructors who had "'little or no knowledge of how to teach, but merely present[ed] to the pupil a series of sequences taught to them at F.I.S.'"[76]

Early in the war, the advanced schools suffered the same problems as the EFTSs. SFTSs were often formed by inducting bush pilots and other experienced civilian flyers into the RCAF and giving them some FIS training. Staff from an established SFTS would frequently be drawn off to form another SFTS, much like the deficient reserve-squadron system of World War I.[77] This dilution of experience at all levels, combined with the geographical dispersion of the schools and the training commands, precluded the RCAF from achieving the standardization envisioned by proponents of the Gosport method,[78] made it almost impossible to inaugurate uniform training methods, and confounded AFHQ policy initiatives designed to impose fixed pass/fail criteria.

At first, RCAF instructors tended toward a lenient approach with their student pilots, as had been the practice in RFC/RAF Canada during World War I. In September 1940, however, a policy of strict testing was instituted by Leckie, who believed that standards were too low. Wastage increased, and eventually it was concluded that the pendulum had swung too far the other way and that the instructors had become too harsh. In September 1942, accordingly, the schools were directed to pass a greater proportion of their students.[79] But since no central authority assumed responsibility for monitoring the implementation of the new policy, standards varied greatly from school to school and from Command to Command, as did wastage rates.[80] Such inconsistencies raised questions about the quality of some Canadian-trained pilots. However, while their skills were examined and found wanting in certain respects,[81] it is perhaps less well known that the RAF had problems of its own producing aircrew of a sufficiently high calibre.

According to the RAF's Air Historical Board (AHB), all training units in World War II followed "a uniform doctrine of [pilot] training" whose origins could be traced back to the Gosport method.[82] Early in the war, however, the quality of instruction was generally poor,[83] owing to the lack of attention paid by the RAF to training development after 1918. To address one element of this problem, a four-week course for civilian EFTS instructors in the United Kingdom was introduced in July 1940.[84] At a conference on aircrew training held in early 1942, delegates from Britain and the Dominions agreed that the quality of instruction in the BCATP needed to be improved as well. Out of this conference came the Empire Central Flying School, created to standardize aircrew training throughout the Commonwealth.[85] But until instructional methods could be monitored in all parts of the training system, the application of a uniform system was problematic. One flight commander at a Canadian SFTS suggested that a number of graduates probably died while on active service because only students with "pretty severe" problems were washed out.[86] As early as 1941 Bott had reported that a "disconcerting number of [BCATP] lads with wings [were] not making good at OTUs here [in the United Kingdom]." He commented further that "they are a heavy loss and will be even harder to dispose of than the large crop of CT [cease training] cases which we already have at Trenton."[87] His remarks were an early indication that Canadian-trained pilots were having difficulty coping with the conditions in Britain.[88]

British Air Ministry inquiries from 1941 to 1943 revealed that the flying abilities of BCATP graduates, especially at night and in poor weather, did not meet the required standard.[89] Yet it was unfair to place the blame for this entirely on the Canadians, for until the end of 1941 the RAF training syllabus, on which the RCAF course was based, concentrated on daytime visual flight procedures and neglected instrument flying. In early 1941 the RAF Directorate of Flying Training issued a report that was severely critical of the instrument-flying training provided by the air force. As an AHB study later remarked:

Instrument flying was neither standardized nor correlated with operational requirements, and ... SFTS instructors were largely ignorant of the first principles of instrument flying and of its importance in operational work. In fact 95 percent of the instructors examined were not noticeably better on instrument flying than the average pupil turned out of a [sic] SFTS.[90]

Thus it is clear that the RAF bore some responsibility for the difficulties faced by BCATP graduates.[91]

Central to the question of the quality of pilot instruction in World War II is the issue of training standards. The British had considered the issue in 1934–36, in response to Tedder's proposed reforms, which embraced higher standards of proficiency and better training methods. However, his proposals for a reassessment of RAF flying-training methods by experts was "dismissed as impracticable" by Air Ministry officials.[92]

Until 1941 training standards "had been largely dictated by necessity," but, beginning in 1942, pilot surpluses allowed training time to be doubled. The RAF's New Deal increased the number of hours of pilot training to 220–270, extending over seventeen-and-a-half to twenty-three months, thus bringing the amount of training up to what was thought to be the German standard. For pilots it also placed more emphasis on night flying and instrument flying, and other aircrew benefited from improved training as well.[93]

While some changes, such as in course length, could be introduced relatively quickly, others, particularly standardization, took longer to realize, especially in Canada. The subject of training standards in the RCAF is still controversial. Most commentators believe, as does one of the authors of the official history of the Canadian air force in World War II, that, despite persistent difficulties in implementing standardization, the bulk of the RCAF's instructional problems had been solved by the end of 1942, and that "by and large" all the pilot graduates of the BCATP were well trained.[94] It is worth noting, however, that investigations conducted in 1943 and 1944 by psychologists at two BCATP flying schools, Pendleton and Arnprior, uncovered some surprising facts about the effectiveness of BCATP training. Both the Arnprior investigation (which was briefly discussed in chapter 2) and the Pendleton investigation were designed primarily to evaluate the effectiveness of the pilot-selection processes, but they also yielded important information about wastage rates during EFTS training. The experimental procedures used were closely controlled by and carefully analyzed at AFHQ. The Pendleton group was composed of pilot candidates selected and trained in the usual way, and it experienced a normal wastage rate, for an EFTS course, of just over 23 per cent. The Arnprior group was treated differently; whereas the Pendleton group excluded candidates who had performed poorly on their initial pilot-aptitude tests, such people were kept in the Arnprior course as a control measure. Surprisingly, the Arnprior wastage rate was an incredibly low 9 per cent.[95]

How was it that a specially selected group could experience more than twice as much attrition in elementary flying training than a group for which the selection standards were much less rigorous (being limited mainly to medical grounds)? According to a report prepared by the AMP branch in August 1944, the discrepancy was mainly attributable to changes in how BCATP flying training was conducted. The Pendleton group completed EFTS in the summer of 1943, the Arnprior group in early 1944. Between those courses, standards flights, similar to those used with the Armour Heights System in World War I, had been added to each EFTS. Previously, all standardization had been carried out by the RCAF's CFS at Trenton.[96] Under the supervision of the new local standards flights, the quality of instruction improved and uniform testing methods were enforced. The AMP report determined that the improvement in training produced by the new standards organization "largely account[ed] for the lower wastage rates found in the Arnprior group."[97] In the words of a narrative prepared after the war by the Directorate of History (DHIST) of the Department of National Defence (DND) (Ottawa), "The shocking difference in the statistical results of the Arnprior and Pendleton experiments ... suggests that EFTS instruction was much worse than necessary during most of the BCATP courses, and that the wastage from these courses was much higher than need have been the case had methods of rejection been more exact, and the standard of training higher."[98]

The improved flying-training scheme was not introduced until April 1944, and thus benefited only the final three of the 106 BCATP pilot classes to graduate in World War II.[99] Had the air force heeded the advice of some of the members of a civilian technocratic élite – psychologists – the changes might have come sooner.

Bott's lectures at the RCAF School of Aviation Medicine in 1940 had included a section on "the psychological aspects of flying training."[100] Other worthwhile material on teaching techniques was also available to BCATP planners. For example, a psychological study of flying instruction presented to the Associate Committee on Aviation Medical Research in June 1941 noted that pass/fail "criteria were essentially a personal opinion," and that "no objective means [existed] to evaluate [the] actual failings of the trainee."[101] Unfortunately for Canada's war effort, there seemed to be a widespread "lack of psychological education" among senior Canadian officers, and little understanding of how psychology could help minimize losses.

Despite the efforts of some Canadian psychologists to bring their discipline to the air force's attention, virtually nothing was done to employ psychological principles to enhance RCAF training.[102] Early in the war, Bott and Banting proposed that the RCAF give flying training to two young psychologists, who would then be able to advise the air force on its training methods. This suggestion was rejected by air force authorities, who believed that pilots were the leading experts on flying training and that the air force had nothing to learn from scientists who had no flying experience.[103] RCAF training thus remained outside the purview of psychological science.[104]

This refusal to accept expert civilian advice may have cost the RCAF many valuable aircrew, as the following two cases illustrate.

In the first case, an RAF MO undergoing pilot training in Canada in 1944 remarked that, even though research had been conducted in Britain on teaching students how to achieve the best round out before landing, "there was a wide divergence in instructors' 'patter'" when they attempted to describe the manoeuvre. Aside from increasing the students' difficulty in learning one of the trickiest parts of elementary flying, this caused "episodes of bad landings," which were often dangerous.[105]

The second case involved Chant's subordinate D.C.Williams. Near the very end of the war the air force finally decided that a psychologist could be accepted into pilot training as a "participant observer." Williams, a young RCAF officer at the time, was the one chosen, and as a result of his experiences he suggested that "cumulative learning would be enhanced" if "regular repetition of new manoeuvres [was] ... standardized as much as possible." His flight commander, however, characterized this idea as "subversive and impractical," confirming Williams's view that the opinions of senior air force officers in charge of training were "not susceptible to significant change."[106] In fact, all Williams was suggesting was the proper implementation of the original Gosport method, brought up to date in accordance with modern principles of psychological research, but by that time distinctive local practices had become firmly established.

As with aircrew selection, only when it was being wound down did the training system reach peak efficiency. Beginning in mid-1943, forecasts of aircrew surpluses caused the BCATP to reduce its output of flyers, which in turn resulted in training-unit staff being released to front-line squadrons. Yet surpluses continued to accumulate, and by February 1944 BCATP flying instruction was at a virtual standstill.[107]

By war's end, the BCATP had overcome most of the formidable obstacles in its path, and had developed into a first-class training system. Besides producing better-prepared flyers, the introduction of a structured, standardized training process gradually changed pilots' attitudes toward flying. In World War I the RFC/RAF had sought to recruit "knights of the air, mounted on Pegasus," but in World War II Bomber Command and Coastal Command – the main users of BCATP aircrew – required a different kind of pilot. Most World War II pilots flew multi-engined giants manned by specialists, where precise instrument flying and meticulous teamwork were essential. As the war dragged on and Bomber Command demanded a greater share of the aircrew pool, the inculcation of professionalism into graduates of the BCATP became more pronounced. While the first BCATP students were not discouraged from engaging in the same kind of stunt flying as their World War I forebears – with equally tragic consequences – later students were subject to stricter air discipline and had impressed on them the notion that flying required more than just "guts."[108] The adoption of standardization and uniform teaching techniques, combined with a new style of training that focused almost entirely on flying skills, created a new breed of military pilot, one who was an aviator first and an officer second.

These changes can be traced back to World War I, when unnecessary casualties compelled the RFC/RAF to develop a more effective training system. The Gosport method was the RFC's solution to the need for better-skilled pilots. After 1918, however, the economic pressures of peacetime resulted in the decline of the training systems in both Britain and Canada. The only way for the training systems to be ready to meet the challenges of another major war was through controlled change; however, for this to occur, they needed regular scrutiny by both military and civilian specialists.[109] The RAF's interwar indifference to training, and Canada's dependence on the British for training doctrine, left the RCAF with a poor model to emulate. The Gosport method was a sound basis to build on, but it was neglected between the wars, and poorly copied in the expansion phases of both air forces. In Canada, moreover, a geographically dispersed training organization and a dearth of standards officers to regulate training norms gave rise, until 1944, to many local variations in the BCATP course of instruction. But whatever their level of training, large numbers of the Plan's graduates were sent overseas to fight in great air battles that claimed the lives of many of them.

The official Canadian air force history has suggested that it is impossible to know whether better training standards could have reduced aircrew casualties. It has been argued that the sheer quantity of aircrew produced by the BCATP in the early days of the program was important in establishing air superiority in the various theatres of war, and in limiting the damage done by enemy forces.[110] This argument is convincing when applied to the Battle of the Atlantic and the North African campaign, but it is less compelling in the context of Bomber Command operations in 1942 and 1943. Britain's official air force historians, writing in the decade after the war, were more critical of Air Ministry policy. They pointed out that Bomber Command was the only operational command that did not have to react directly to enemy initiatives, for it had the luxury of reducing losses by limiting its sortie rate.[111] Many of the pressures to expand the size and the role of Bomber Command came from within the RAF itself, and could have been controlled by the service.[112] The critical OTU situation in 1941–42, for example, compelled the air force to modify operations to conform to training capacity. Moreover, in early 1942 the RAF's CAS believed that the scale of the attack that Bomber Command would eventually mount would depend more on attrition than on any other factor, and he was prepared to reduce wastage due to inadequate training by diverting up to 70 per cent of operational aircraft to the OTUs.[113] As was suggested in chapter 1, perhaps a more deliberate and limited build up of Bomber Command, with more attention to the quality rather than the quantity of aircrew produced, would have reduced losses and improved the efficacy of both training and operations.[114]

In Canada, however, other problems adversely affected the country's contribution to the air war. In contrast to the way in which selection issues were handled, the RCAF eschewed the advice of technocratic élites concerning its aircrew-training system. The British, on the other hand, came to trust the advice of scientists, eventually establishing a Training Research section that included two Canadian psychologists as advisers to the AMT.[115] It is ironic that Bott, Canada's pre-eminent psychologist, left his consulting role with the RCAF to advise the British air force on matters that were soon to become crucial to the success of Canada's air force. If civilian experts in Canada had been invited to perform a function similar to Bott's in Britain, it is possible that the RCAF would have been able to improve its training process sooner.[116] As for the question of autonomy, the RCAF's reliance on British pilot-training methods left it vulnerable to the weaknesses inherent in that system. With no institutional means of preserving the lessons of the past, the RCAF was condemned to repeat the errors of its World War I predecessors.

4 A Predisposition to Cowardice? Aviation Psychology, 1914–45

In both world wars, psychological casualties were responsible for a huge drain on British and Canadian air force manpower. In 1944, for example, it was found that "mental disorders" accounted for one in three medical discharges from the Canadian air force.[1] But with regard to aircrew lost on operations it was often difficult to know whether psychological factors had contributed to their deaths. The causes of most combat losses were never precisely established, and in many cases it was difficult to know whether carelessness, fatigue, or impaired mental function produced by combat stress[2] contributed to an aircraft's damage or destruction.

Uncertainty about the causes of fatal errors made by military aviators dated back to World War I. One RFC MO of that period observed that aviators suffering from "flying fatigue" (exhaustion caused by combat stress) showed the same outward symptoms as those who were simply tired. Whatever the cause of his exhaustion, on meeting an enemy the tired pilot was often lethargic, could not "'trouble to think about manoeuvring,'" and in this way often met his end.[3] This was just one kind of error, but of course there were many others. Studies conducted by the RAF in World War II revealed that "human error" accounted for 70 per cent to 80 per cent of all aircraft accidents,[4] and an interwar study of pilots "in the United States Air Service" found that about 80 per cent of groundings of qualified pilots were due to "nervous instability."[5]

Whatever the exact numbers of airmen lost for psychological reasons, by the end of World War I it was accepted that, in combat conditions, the abilities of large numbers of flyers could be impaired by their psychological state. The causes of impairment were diverse, the symptoms numerous and often difficult to differentiate from those of other ailments. Some of the psychological problems experienced by aviators resulted from flying itself, others from combat, and some, of course, were unrelated to military service. Treatment, the medical authorities of the belligerents agreed, should be simple and administered near the front lines to be most effective.

As was the case with scientific findings concerning aircrew selection and training, early discoveries in the psychology of air warfare reached Canada by way of the RFC/RAF, for during World War I Canada's air force medical service depended on British research in aviation medicine. Similarly, the RCAF deferred to RAF experts during the interwar years and early in World War II. To understand the RCAF's treatment of psychological casualties in 1939–45, it is therefore necessary to know something about the evolution of aviation psychology in the RAF.

Aviation psychology today encompasses many subjects, including aircrew selection, human factors research, leadership, performance assessment, and clinical applications.[6] The discussion here, however, is principally concerned with the development of clinical approaches to the diagnosis and treatment of aircrew suffering from combat stress. As we shall see, aviation psychology made a crucial contribution to the RCAF's efforts to solve the manpower problems associated with air warfare in 1939–45.

Strain due to flying, as was mentioned above, was not unique to World War II. When men first took to the air, the sheer effort involved in controlling poorly designed, temperamental aircraft exacted its toll on aviators both physically and mentally. An early commentator on the psychology of flying remarked that "the intense nervous strain of flying will, in a short time, so wreck the strongest nerves as to render a man unfit to go aloft ... In this suggestion a probable explanation of some of the apparently inexplicable accidents of aerial navigators is found."[7] As pioneer aeronauts accumulated flying experience, the stresses induced by the novelty of flight and by the demands of managing primitive vehicles diminished, but improved aircraft performance placed other strains on aviators. The effects of cold, anoxia (lack of oxygen), G forces, and unusual aircraft attitudes unrelentingly exacted their own mental toll, and the advent of war imposed the ultimate stressor – combat – on the fledgling flyers.

The psychological problems of flight were first studied by nonspecialist medical doctors with an interest in aeronautics. Some work on aviation psychology was done by the British during the early years of World War I, but with no aircraft available for research, and no separate RAF organization to encourage and coordinate the efforts of investigators, little experimentation was possible.[8] Nevertheless, by war's end a body of psychological knowledge, based on practical experience in the field, was beginning to emerge.

In World War I it was understood that aviators faced risks different from those experienced by soldiers in the trenches. It was hypothesized that aviators were more susceptible to mental breakdown because, unlike the junior infantry officer who performed his duties "on the ground and in a crowd," flyers often worked alone, and were under intense psychological pressure every time they went aloft, whether to do battle or for a simple air test.[9] The extreme contrasts of a military aviator's life were thought to intensify the strain. His days consisted of "long spells of idleness punctuated by moments of intense fear," and fear was the "most intense strain to which the human nervous system [could] be subjected."[10]

Consequently, as the war progressed pilot wastage attributable to mental strain reached serious proportions among air forces,[11] and forced the military authorities to conduct urgent inquiries in an attempt to check the losses. These inquiries bore some fruit, and by the end of the war a consensus had been reached on many of the major issues in military aviation psychology. First of all, most experts agreed that the psychological aspects of flying were at least as important, if not more important, than the physical aspects.[12] Second, they concluded that while improved selection procedures had substantially reduced wastage attributed to psychological factors, as the war neared its end psychological losses began to mount again as the quality of recruits declined.[13] Third, it was generally recognized that mental breakdowns among aviators were chiefly caused by fear, and that fear – "getting the wind up" – was "a normal reaction to a very abnormal environment,"[14] namely the arena of air combat. Fear and other combat-related stressors, psychologists agreed, could lead to weight loss, tremors, abuse of alcohol and tobacco, insomnia, nightmares, cardiac irregularities, loss of confidence, and general nervous breakdown.[15] Fourth, investigators observed that the types of stress different aviators faced influenced how long they could function effectively. For instance, it was determined that fighter pilots suffered more stress than other flyers – and thus more breakdowns – because they often flew alone and had to contend with the unpredictable actions of their adversaries. It was noted, too, that living conditions and squadron morale affected resistance to breakdown, and that some flyers handled stress better than others.[16] Finally, specialists concluded that the best treatment for combat stress was to talk about one's problems with a sympathetic MO and to look to simple medical remedies to relive physical symptoms. For those who had been in combat for prolonged periods, rest while on leave away from the front completed the cure.[17] Other, more complex regimes, including Freudian psychotherapy, showed little promise.[18]

Postwar analyses based on the wartime experience of all the Allied fighting services confirmed these findings. It was noted that for those who broke down after limited exposure to combat, recovery near the front, "or at least 'somewhere in France'" was "more prompt and durable" than treatment away from the front, and that evacuation to hospitals in England often turned the "temporarily incapacitated" combatant into a "chronic invalid."[19] Observation also revealed the importance of cumulative stress in precipitating breakdowns; a family crisis on top of the pressures of life at the front could trigger a collapse.[20] It was believed, as well, that a lack of appropriate diagnostic and treatment resources forced the armed services to discharge large numbers of men, most of whom became chronic invalids, although some apparently recovered once removed from the stresses of battle.[21]

A leading figure in the postwar study of military aviation psychology was James L. Birley, a prominent British neurologist who had served with the RFC/RAF in France from 1916 to 1919.[22] Birley suggested that combat aviators passed through three stages of psychological reaction to their environment, which he called the phases of "inexperience," "experience," and "stress." He noted that 70 per cent of the casualties among RFC/RAF flyers occurred during the phase of inexperience, especially the first three months of active service. The next phase, experience, marked the "zenith" of the aviator's effectiveness. This was almost inevitably followed by the third phase, stress, when the flyer became "stale" or burnt-out. Birley believed it was the MO's job to prolong the second phase as long as possible by resting the aviator "at the critical moment" before staleness or burnout occurred. Although it was easier to arrange treatment for a flyer than for an infantry officer – for the reason that it was less disruptive to military organization to send the former away on leave for a rest – the flying officer in his solitude was more susceptible to mental collapse because he "lacked the inspiration born of responsibility to his men" and the "comradeship in danger which for the infantry served as sources of support and encouragement."[23]

Birley's observations owed a great deal to his close association with aviators. He was described by one squadron commander as "a brilliant nerve specialist, who devoted an immense amount of energy to his work for the RFC."[24] This included visiting and socializing with front-line flyers to gain firsthand knowledge of their problems. What Birley found was that virtually every aviator who remained in action long enough became a psychological casualty. Accounts of those who flew in World War I bear this out; almost every aviator who recorded his ordeals at the front admitted that he became stale after exposure to too much stress.[25] Recognizing the inevitable effects of stress on their charges, most squadron commanding officers (COs) dealt with the problem in what they believed was the humane way, by posting the affected aviators back to Britain.[26]

Birley and other contemporary observers remarked how the war created special pressures for front-line leaders. Stress could permeate a whole squadron if its leaders, especially its flight commanders, suffered burn-out. Problems also arose when squadron commanders, understandably reluctant to lose their most experienced aviators, allowed them to become stale.[27] Manpower shortages in the latter part of the war created further difficulties. As casualties mounted in the RFC, squadron commands passed from older, more mature flyers to men in their early twenties,[28] who favoured an aggressive command style and tended to lead by example.[29] This kind of leadership, however, had its drawbacks. By April 1917 losses among squadron commanders had become so great that they were forbidden to fly within five miles of enemy lines. This prompted one CO to return to the cavalry rather than command a squadron he could not lead in the air. He would not "command a slaughter" if he was not permitted to share the risks of his subordinates.[30] By the end of the war the RAF faced an acute shortage of experienced leaders, which forced the flying service to adopt the expedient of giving flying training and squadron commands to some senior ground officers "skilled in the handling of men."[31] It is hard to imagine how these commanders could have earned the respect of their juniors when they had never been in air combat and were prevented from winning the confidence of their men by fighting alongside them in the air.

Another doctor who investigated the psychological aspects of military aviation was Martin Flack, a senior RFC/RAF researcher who became Director of Research for the RAF Medical Service after the war.[32] A civilian physiologist before the war, Flack nonetheless believed that the performance of military flyers was determined as much by psychological as by somatic factors. His "U tube test" attempted to assess both by having the subject raise a column of mercury by blowing into a tube. This was designed to measure lung capacity and expiratory force, and since it had been observed that flying stress rendered individuals physically as well as mentally tired, Flack hoped to identify cases of staleness in their early stages. He believed that his approach was more objective than other tests, and he also argued that it allowed an aviator's performance to be compared over time.[33] The U tube test was also employed in World War II, although by mid-1941 both the RAF and the RCAF were beginning to question its usefulness. Its use was gradually phased out by both services before the end of the war.[34]

The work of Birley, Flack, and other senior researchers in aviation psychology reached a wide audience through the *Lancet* and other medical journals. Birley's findings were laid out in a series of lectures to the Royal College of Physicians of London, which were published in the *Lancet* in 1920. H. Graeme Anderson, a Royal Navy surgeon, published *The Medical and Surgical Aspects of Aviation* in 1918. This book was the first to deal with aviation psychology in a comprehensive way, and as late as 1942 the United States National Research Council called it a "valuable" reference for American flight surgeons (although by then it was out of print).[35]

By the end of World War I, then, the study of the military side of aviation psychology had emerged as a full-fledged discipline. However, little progress was made between the wars. Because financial cutbacks dictated that hardly any original research could be conducted, most of the literature that appeared was based on information gathered in 1914–18. Birley remained a leader in the field of aviation psychology in the 1920s, and in a lecture at the RAF Staff College in 1923, later published in the *Lancet*, he re-emphasized his conclusions that Freud's theories, although useful in some circumstances, were "open to important criticism" when applied to "normal behaviour," and that a person's ability to control fear was weakened in war by stress-related fatigue. In Birley's opinion, it was the service's responsibility to adequately prepare its men for combat by applying psychological principles derived from wartime experience.[36]

Some psychologists disagreed with Birley, among them Frederick C. Bartlett, the author of *Psychology and the Soldier* (1927), one of the few interwar books on combat stress. Bartlett's work was written to acquaint professional soldiers with the practical applications of psychological research, and it addressed many of the issues that were to become crucial in World War II. A founding member, in 1939, of the Flying Personnel Research Committee (FPRC), Bartlett was one of Britain's most distinguished psychologists.[37] His book summarized much of the work that had been done during World War I, but its major contribution was its comprehensive application of some aspects of Freudian theory to the psychological disorders of war. Bartlett argued that success or failure in adapting to military life was determined mainly by one's temperament. He proposed that "weak-lings" (mentally unfit personnel) should not be allowed into the armed services,[38] and that only men with the right kinds of tem-perament, the kinds that gave men the strength of will to resist mental breakdown, should be kept on. Bartlett added, however, that even among men carefully screened for military service, breakdowns similar to those encountered during World War I would occur be-cause of individual differences in temperament. Conversion hys-teria,[39] for example, occurred less frequently among officers than among the other ranks, because the temperament of the private sol-dier did not allow him to "very vigorously fight against his fears."[40] Officers, on the other hand, suffered more than their men from anxiety neurosis and consequent general collapse, because their tem-perament predisposed them to struggle "strenuously against an ad-vancing mental breakdown."[41] Bartlett added that fighting men who suffered "mental breakdown nearly always [had] a long history" of psychological problems, no matter what their rank.[42] Despite his use of much of the information gathered by the British Army during the war, he paid little heed to the conclusions of specialists about the utility of Freud's therapeutic method. Curiously, Bartlett's proposed treatment regime for psychological casualties in the field envisioned psychotherapy by junior officers, who would employ Freudian tech-niques of dream analysis and "free association."[43] Bartlett rejected the behaviourist view advanced by an American, J.B. Watson, that fear was essentially a learned response that could later be unlearned. According to Bartlett, negative experiences in the early years of life profoundly affected the psyche, and the fears that originated from those experiences could usually be overcome only by psychotherapy and a "tremendous effort" of will on the part of the patient.[44]

Unlike their British counterparts, American psychologists were unable to complete much research on aviation psychology and com-bat stress during World War I because of their country's late entry into the conflict.[45] However, their postwar syntheses of existing data provided some of the best insights into the challenges of wartime psychology. One of the most valuable American works on the sub-ject, published in 1929, was a volume on neuropsychiatry[46] in the United States Army's official medical history of World War I. Based for the most part on British experience in the field, it was described in 1943 by the consulting psychiatrist to the British Army as "a complete textbook on both the clinical and administrative sides" of military psychiatry.[47] The volume's authors reported, among other things, that the American, British, and French medical services had all found that simple treatment, administered quickly and near the front, improved the chances of recovery of those suffering combat-related nervous problems. They concluded, moreover, that no new types of mental disease had been observed in soldiers and that there were no distinct "war psychoses." Relying on their understanding of Freud's theory of defence mechanisms, they also argued that nervous disorders were often physical manifestations of unconscious conflicts, and therefore beyond the patient's control.[48] Another important point was that psychological casualties were a "medico-military problem," one that required doctors-in-uniform to place military priorities ahead of their Hippocratic duty to individual patients. This meant that when manpower needs were "seriously threatened" by the psychological afflictions of war, the aim of the medical services should be to return men to combat as soon as possible, even if such a policy was not in the best interests of the individual fighting man.[49] In any event, it had been found during the war, to the surprise of many specialists, that soldiers actually recovered their mental equilibrium more quickly under a treatment regime that eschewed long-term intervention.

Five years after the publication of the American volume, major changes occurred in RAF psychological medicine. The death, in 1934, of Birley and Flack, two of the RAF's most experienced specialists in aviation medicine, opened up the field of military aviation psychology in Britain to people with formal training in the study of human behaviour. Whereas the field had been dominated by RAF MOs with wartime experience, it now began to come under the sway of psychiatrists and neuropsychiatrists,[50] whose explanations of mental breakdown focused on the inadequacies of the individual aviator, much the same as Bartlett's had. These ideas were a significant departure from existing RAF doctrine as enunciated by Birley.

By 1939 the new theories had become the next RAF orthodoxy, and were reflected in the first edition of "Notes for Medical Officers on the Psychological Care of Flying Personnel," which was issued as Air Ministry Pamphlet 100 (AMP 100) in May 1939. This document drew on the experiences of World War I doctors when it described the symptoms that had been observed in psychological cases, and some of the treatments that had been used,[51] but where it differed strikingly from previous air force medical opinion was, first of all, in its reliance on an interpretation of Freudian theory to explain the behaviour of patients, and secondly, in its endorsement of treatment by neuropsychiatrists.[52] Although AMP 100 acknowledged that stress was cumulative and that "everyone has a breaking point," it emphasized the role of character defects in most cases of mental breakdown. Psychological theory, as outlined in the pamphlet, held that the "instinctive tendencies" in man were amenable to control by the higher levels of the mind.[53] In combat, the most basic human instinct – self-preservation – which often manifested itself as an urge to flee from danger, could be suppressed by controls acquired "during education and training," which comprised one's "character." Those with "strong characters" were able to display "patriotism," "tenacity of purpose," and self-sacrifice, while those with "weak characters" were described as "vacillating," "undependable," and "ineffective."[54]

The pamphlet recommended many of the prophylactic measures that had been used in World War I to preserve the mental health of airmen, including physical exercise programs, periods of rest and leave, and the fostering of esprit de corps. With respect to treatment the advice was also similar to earlier practice, except for the recommendation that MOs provide basic psychotherapy, which had been labelled unsuccessful by most British practitioners in World War I.[55]

The most significant concept put forward by this publication was that individuals who suffered psychological breakdown were largely the authors of their own misfortune. Their "weak characters" made them more susceptible to the stresses of war, and therefore unfit to serve in the air force. Unlike a physical disability, their psychological complaint was a result of their own inability to control their fear. An aviator with shattered nerves might be salvaged by "a little commonsense psychotherapy" – though what exactly that meant was not explained – but the "result [would still] depend very largely on the stuff of which he [was] made."[56]

AMP 100 reflected majority opinion among British psychologists and psychiatrists on the eve of World War II. In the 1930s hereditarianism dominated the behavioural sciences, and it was believed that intelligence was largely an inherited trait that could be measured fairly accurately. It was also accepted that there were solid links between intelligence and leadership potential.[57] At the same time, Freud's theories, which focused on the effects of childhood experiences, were highly influential, and formed the basis for the clinical interpretation of many mental disorders.[58] In combination, hereditarian and Freudian concepts supported the view that mental breakdowns were the product of innate characteristics and that some people were predisposed to collapse. "Breeding" and one's experiences during childhood and early adolescence were believed to determine one's chances of forming such a predisposition, and many students of human behaviour in military settings rejected the notion that military training could eliminate most tendencies toward breakdown.[59] With war imminent, British specialists accordingly proposed that psychiatric and other tests be used to exclude "weak" candidates for military service.

Once the war began, the Air Ministry, having concluded that existing knowledge about aircrew psychological disorders was insufficient, initiated several investigations of the subject, which resulted in a series of reports for the FPRC. Most of these reports – over thirty in number – were authored or co-authored by three members of the RAF Medical Branch, all former civilian practitioners who were sent on active service in 1939. The most influential of the three was the senior RAF consultant in neuropsychiatry, Group Captain (later Air Vice-Marshal) Charles P. Symonds.[60] Symonds, more than any other specialist, defined RAF medical policy on combat stress. In some of his work he was assisted by a young neuropsychiatrist, Squadron Leader (later Wing Commander) Denis J. Williams.[61] The third officer, Squadron Leader Donald D. Reid, a Bomber Command MO who later worked with the Director General of Medical Services (DGMS) at the Air Ministry, produced a number of field studies and statistical analyses that complemented Symond's work.[62]

Beginning in 1939, RAF specialists in psychiatry and neurology, and flying personnel MOs,[63] visited operational stations "to investigate neuropsychiatric problems" and "flying stress." They found little evidence of such conditions, but they warned that intensified operations might very well bring forth psychological casualties.[64] And this prediction was soon proved true.

One of Symond's goals was to standardize the terminology and diagnostic criteria employed by specialists in aviation psychology. In a report written in 1941 he outlined the classification system that was to be used for the diagnosis of psychological disorders among RAF aviators. He advised that, except for "fatigue syndrome," MOs should use the same terminology as civilian psychiatrists.[65] Symonds noted that the RAF had made "a concession ... to popular prejudice" by avoiding the psychiatric term "neurasthenia"[66] in favour of fatigue syndrome." The latter had been introduced, according to Symonds, because of a prejudice by senior officers against attaching the label "neurotic" to any man who had "done his best."[67] Symonds also insisted that "flying stress" was not itself a disorder, and that the term should only be applied to the *causes* of psychological disorders among airmen.

In an attempt to distil the RAF Medical Branch's understanding of psychological disorders experienced by aircrew, Symonds and Williams prepared two seminal reports for the FPRC in early 1942. In the first they reviewed the existing literature on psychological disorders among airmen. They concluded, among other things, that the field of aviation psychology had not been properly studied because it had "escaped the notice of trained psychiatrists" and had been left mainly to non-specialist MOs. Despite their concerns about the state of the literature, Symonds and Williams's recommendations were similar to those of previous observers in the field. The literature, they found, identified fatigue as the "most important single predisposing cause of psychological breakdown in flying personnel," and fear as the main cause of fatigue. While noting the impact of in-flight physical factors on the mental health of aviators, they stated that "of particular importance is individual predisposition to fear, [which is] largely dependent upon temperament." They maintained, as had Birley, that it was the MO's duty to rest aircrew "at the crucial moment" before breakdown, so that the aircrew could recover from their fatigue and be returned to duty in a short time. Symonds and Williams also observed that physical fitness and the promotion of group morale could reduce the effects of fatigue. Apart from this, they held out the hope that better selection methods would permit the identification of aircrew temperamentally unsuited for military operations.[68]

The second FPRC report by Symonds and Williams, which drew on the observations of the RAF's specialists in aviation psychology as well as its non-specialist MOs, summarized what was known about aviation psychology by March 1942. The report expressed concern that the terminology problem had not yet been solved, and that there was still "great confusion of thought," which was apparent from the imprecise vocabulary used by "those concerned with the psychological welfare of aircrews in the RAF."[69] With regard to diagnosis and treatment, most of the doctors cited in the report followed prevailing opinion in holding that it was largely the airman's character that determined his resistance to the stress of flying.[70] The report included such statements as "'more intelligent members of aircrews ... recover completely with rest and encouragement'" and "'a man with a good family history, who has never suffered from any nervous disability, develops war neurosis only under very exceptional circumstances.'" One psychiatrist went so far as to label some airmen who suffered breakdowns "'constitutionally timid.'"[71] Although Symonds and Williams did discuss the recognized strategies for staving off the appearance of disorders among aircrew, including the granting of leave, limits on operational tours, provision of alternatives to combat duty for weary flyers, and making recreational facilities and other amenities available to aircrew, the general tone of the report was reflected in a remark by the senior MO of Flying Training Command, who said that the "important task for psychologists is not to patch up inferior material which has failed, but to eliminate it before it can fail."[72]

This attitude was particularly noticeable among the Bomber Command doctors quoted in the report. Air Commodore F.N.B. Smartt, Bomber Command's Principal Medical Officer, was quoted as writing, in January 1941, that "'the importance of temperamental unsuitability in causing psychological disorders in members of air crew'" required the "'radical elimination of those unsuitable individuals.'"[73] Squadron Leader Reid, who was still with Bomber Command at the time of the report, advocated the removal of "aircrew lacking in moral fibre" because of their "bad influence," and stated that "firm treatment with degrading [meaning loss of brevet and rank] and re-posting [was] very important."[74] While the Bomber Command medical advisers recognized that the stresses of OTU training and operational flying could cause fatigue in "normal" flyers, they held that there were men who, by virtue of their genes and upbringing, were likely to break down under stress and exhibit neurotic symptoms.[75] Such men were unfit to be aircrew, and were therefore to be treated without sympathy.

This severe attitude can perhaps be partly explained by the number of psychological casualties Bomber Command was suffering at the time. Although a comparison of "flying stress" casualty rates among Bomber Command, Coastal Command, and Fighter Command is an uncertain undertaking, given the terminological confusion mentioned earlier, it is worth noting that after one year of war Bomber Command reported an incidence of "'Flying Stress'" of 2 per cent, compared to 1.4 per cent for Coastal Command and 1.2 per cent for Fighter Command.[76] By January 1941 the Bomber Command rate had increased to "an annual total of 5 percent of operational air crews [who] developed psychological illness of sufficient severity to lead to admission to hospital or disposal by CME [the RAF's Central Medical Establishment]." Because the 5 per cent figure included only officially recognized cases, Air Commodore Smartt concluded that "the total number of psychological disorders in aircrew ... must be very much higher."[77] Bomber Command's mounting psychological losses represented a very worrying situation for a fighting force that was already suffering severe manpower problems caused by expansion and operational attrition.

By the end of the war, published British medical opinion was quite consistent in its approval of the views of Symonds, Williams, and Reid, who had continued to produce their influential reports.[78] Lord Moran, Churchill's personal physician and the author of *The Anatomy of Courage*, considered by some a classic study of the fighting man, asserted that war cannot "turn men of sound stock into cowards ... it was only bad stock that brought defeat."[79] Similarly, a book published by two RAF doctors in 1945 declared that the paramountcy of predisposition in cases of mental breakdown was "one of the most important lessons learned by psychiatrists during the war."[80] For support the RAF doctors could point to a report published by Williams in February 1943 which estimated that approximately 22 per cent of RAF aircrew were predisposed to psychological disorders. While Williams admitted that it would be "uneconomical" to remove all these flyers from combat duty, he suggested getting rid of the "severely predisposed individuals," who by his count made up 3 per cent of the total aircrew population.[81]

From a scientific point of view, of course, the data had one crucial defect. They were almost exclusively based on the cases of airmen who had been diagnosed as suffering from a psychological disorder, and no concerted effort was made to examine the temperament of those who had shown no symptoms of mental illness. While it might have been true that "two-thirds of individuals who failed to withstand the stress of flying were predisposed to nervous breakdown," as one wartime report concluded,[82] very little research was done to discover how many of the predisposed *did not* break down.[83] It could have been the case that many of the predisposed survived the psychological rigours of combat and did admirable work. (In fact, controlled studies since World War II have concluded that there are no personality traits that create a predisposition toward mental breakdown in combat.[84]) To avoid this thorny issue, some specialists hid behind what they called the validity of personal experience.[85] They acknowledged that their sample populations were small and that no control groups were employed in their studies, but they justified their conclusions on the basis of their extensive professional training and clinical experience.

After 1945, however, a chorus of criticism was directed against the methodology employed by World War II military specialists in psychology, including those who specialized in aviation applications. These assessments called into doubt the very foundations of psychiatric practice. The remarks of F.C.R. Chalke, a psychiatrist who had served with the RCAMC during the war, were typical. Chalke noted that, in the psychiatric screening of recruits, "the examiner usually felt obliged ... to make a 'diagnosis.'" Professional training and custom dictated that if a person presented himself or herself to a psychiatrist, something must be wrong, and the psychiatrist obliged with a "diagnosis." Chalke's report also questioned some of the fundamental assumptions of conventional psychiatry. He stated that, as a result of his wartime experiences, he "was painfully aware of [the] present lack of basic scientific information" on such important subjects as the relationships between physiological and psychological variables; changes in test scores after diagnosis of illness; what comprised the "'constitutional' components of the tendency to develop mental illness"; whether "'ego strength' [was] a definable, measurable, useful, construct"; and, with respect to patients predisposed to mental breakdown, how to distinguish between those who were able to cope with stressful situations and those who would not.[86]

Some American air force physicians also criticized the approach of the civilian "psychiatrist-in-uniform" because the specialist often believed that, as in peacetime practice, severe symptoms "made for a bad prognosis." Major Douglas Bond, the Director of Psychiatry in the Eighth Air Force's Central Medical Establishment, observed, however, that most young flyers had "'strong recuperative powers'" and that simple treatments such as rest might lead to recovery from "'gross psychic trauma.'"[87] It is easy to see why some military officers doubted civilian psychiatric judgment, when civilian psychiatrists-in-uniform claimed, among other things, that only "a hair divides the normal from the neurotic, the adaptive from the nonadaptive,"[88] and that "[f]lying is rarely an arduous occupation involving much hardship."[89] The doubts expressed by Chalke and by some American air force doctors lead us to the vexing question of the accuracy of clinical assessment in determining predisposition to breakdown.

World War I specialists had agreed that assessing flying temperament was a near-impossible task.[90] But World War II neuropsychiatrists, armed with their own version of Freudian theory, believed that they could succeed where others had failed, and an FPRC study published in June 1943 appeared to confirm their convictions. It concluded that the "prognostic opinions of service psychiatrists are consistent in the degree of accuracy," and whether the person returned to duty or not depended "on the calibre of the human material."[91] These conclusions, however, were called into question by another report, published in October 1944, which looked at a sample of 5,000 aircrew who had been referred to neuropsychiatrists "with suspicion of a neurosis." While it found that clinical examinations were useful in "giving the broad picture" of a population of patients, "disturbing disagreements" between psychiatrists meant that there was a real risk that an airman's problems could be incorrectly diagnosed, resulting in discharge from the service without just cause.[92] Another problem was that large-scale psychiatric screening of aviators was impractical because of the small number of psychiatrists available and the long time it took to conduct a clinical assessment. The result was that initial diagnosis and disposal often fell to the local MO or even to the executive branch. This issue will be discussed further in the next chapter.

As more and more Canadians were sent into air combat after 1939, the RCAF found itself drawn into the debate concerning the psychiatric assessment of airmen. Unfortunately, its meagre medical resources (compared to Britain and the United States) forced it to depend on the experience of others in what was, as we have just seen, a controversial area.

Between the wars, the Canadian air force was served by a tiny army medical corps establishment that could cope only with routine matters.[93] Overwhelmed by the RCAF expansion of 1939–40, the RCAMC handed over responsibility for air force medical matters to a new RCAF Medical Service in September 1940. The first director of the RCAF Medical Service was Group Captain (later Air Commodore) R.W. Ryan, an RAF specialist in aviation medicine on loan to the RCAF since February 1940. He was succeeded, in early 1943, by Air Commodore J.W. Tice, a former RCAMC major who had been seconded to the RCAF in 1939.[94] With the RCAF finally in control of its own medical branch – a service that grew from 258 MOs in July 1941 to a wartime peak of 722 MOs in January 1944[95] – the air force was free to set its own medical priorities. At first, this meant focusing on problems of recruiting and selection, and so it was not until late 1942 that problems of general medical care – including psychiatric care – attracted close attention.[96]

As early as 1936, when the RCAF was just beginning to expand, it was observed that junior officers who spent a lot of time in the air had to be grounded because of "the extreme physical strain of active flying." In imitation of the British approach, short-service commissions of four years' duration were introduced in 1939 to create a pool of young men whose principal duty would be flying. The intention was to relieve permanent air force officers from continuous flight duty, and to rest them periodically by giving them technical and administrative posts.[97] Thus, even before the war, the RCAF was alert to the fact that intense flying could cause burn-out in young aviators. Burn-out increased dramatically following the outbreak of war, and soon became an important concern for the RCAF Medical Branch. The RCAF, however, lacked expertise in aviation psychology, and for this reason had to rely on civilian experts to train its many new MOs.

The RCAF, indeed, even had trouble locating published material on aviation psychology. It eventually adopted as its standard medical text the American Harry Armstrong's *Principles and Practice of Aviation Medicine* (1939), the only recent book-length treatment of aviation medicine that was available.[98] The problem was that Armstrong had little to offer on many aspects of aviation psychology. His book concentrated on civilian flying and peacetime military flying, and its comments on wartime operations could be misleading. Armstrong suggested, for example, that owing to the "careful methods of selection" employed by air forces, it was not expected that "serious conditions" of mental breakdown would occur in war, except in rare individual cases. At the same time, however, he admitted that the psychological qualities of a successful aviator had not been adequately defined, and that medical examiners needed to rely on their "experience" and "insight" in judging aircrew fitness.[99] Armstrong's material on psychological standards for aircrew selection relied on USAAC practice, even though the USAAC's Neely Mashburn, head of the Department of Psychology at its School of Aviation Medicine, condemned its standards as arbitrary and subjective, and called the poor flying-training results achieved by the USAAC "an indictment" of the "present method of selection." Mashburn suggested that USAAC procedures were only marginally better than those that could be achieved with an aircrew-selection system based on mere chance. The only reason that the USAAC retained its system, Mashburn added, was that nothing better was available.[100] Clearly, the RCAF's chosen reference work contained some serious flaws.[101]

Later in the war, the RCAF obtained translations of a Luftwaffe pamphlet and of a Russian book on aviation medicine, but they added little to what was already available on aviation psychology from other sources. The German pamphlet contained a very short and general section on how an MO should treat aviators who were "flown out," and stressed the importance of heredity and "natural disposition" as determinants of aircrew temperament. The Russian book, a production of the Pavlov Institute of Aviation Medicine, offered a brief treatment of aircrew psychology that substantially agreed with the RAF's views on the role of temperament, although it supported the view that proper training could overcome deficiencies in aviation students' dispositions.[102]

The RCAF also drew on resources closer to home. Early courses at its School of Aviation Medicine included some Canadian content in psychology in the form of a short presentation by Bott on the "Psychological Aspects of the Maintenance of Personnel." Bott appears not to have strayed much from the published American literature, except for some brief remarks on the role of the group and of traditions and leadership in sustaining morale.[103] Later in the war, a one-week refresher course for MOs at the School of Aviation Medicine included material on "neurological and psychiatric problems."[104]

The presentations at the School of Aviation Medicine notwithstanding, there was always some trepidation about discussing medical problems involving the mind. In early 1941 a junior RCAF MO was asked to give a lecture on psychiatric aspects of pilot selection and training to staff at the Toronto Psychiatric Hospital. Group Captain Ryan refused permission, stating that "the less said about this the better."[105] As late as August 1941, MOs were not asked any questions concerning aviation psychology on their final exam at the School of Aviation Medicine.[106] It appears that senior air force physicians based in Canada were apprehensive about delving too deeply into the pathologies of the mind.

RCAF MOs overseas, however, had to confront such pathologies in a more direct way, and the knowledge they gained was made available to senior staff in Canada. In 1940, for instance, the RCAF received reports on the British experience in handling psychological problems from Lieutenant-Colonel K.A. Hunter, an army officer who was the Canadian air force's senior doctor overseas. He reported that the RAF was instructing its physicians to refer to AMP 100 for the treatment of combat-stress casualties, and squadron MOs were being advised to keep confidential records on the physical and psychological health of every aviator. The RAF, he noted, also considered it important that its MOs become personally acquainted with the aircrew who were their responsibility, and establish good working relationships with squadron and flight commanders (to facilitate confidential discussions about the psychological care of airmen).[107] Another informant, the MO of 110 (AC) Squadron, a Canadian unit, reported that these and other routine tasks kept him so busy that he did not have time to take any professional development courses, including one on the psychological care of flying personnel.[108] Even though RCAF MOs in the field were available to care for aircrew in their units, treatment policies were the responsibility of British medical authorities.[109] Hunter, alert to the potential political and medical problems of this situation, reminded Canadian authorities of the similar problems that had been outlined in the "Bruce report" of World War I, and suggested that adequate Canadian medical resources be deployed to Britain for the care of RCAF casualties, to ensure that the kind of public outcry over the treatment of Canadian wounded that had been heard in the earlier war would be avoided. For the same reason, he also thought it important that all RCAF squadrons have Canadian MOs.[110] Despite Hunter's recommendations, and the similar experience of the Canadian Corps in the 1914–18 war, no separate RCAF hospital service was ever established in Britain. Like their predecessors who were cared for in Volunteer Aid Detachments (VAD) during the first war, many wounded Canadian aircrew were, as a result, "separated from Canadian jurisdiction."[111]

Other information from overseas came to the RCAF when Tice visited Britain from April to July of 1941. He reported that the RAF had at first experienced fewer casualties from flying stress than expected, but that psychological cases proliferated during Dunkirk and the Battle of Britain, when "too few pilots were fighting too frequently for too long." Many of the flyers had been traumatized from being in, or witnessing, a flying accident. The "solution," he concluded was treatment as laid down in AMP 100.[112]

In Canada, the advice of civilian psychiatrists in these matters was not sought by the air force in any meaningful way until the spring of 1943, when a reorganization of the RCAF Directorate of Medical Services opened up a formal structure for receiving advice from neuropsychiatrists. One of their "immediate objectives" was "[t]o review and modify the concept of 'lack of moral fibre'" (an issue that will be discussed in greater detail in chapter 6).[113] The LMF issue had come to the fore when "it became apparent that RAF medical standards and attitudes" were different "in many respects from those of the [RCAF]." To address this problem quickly, RCAF medical staff began to process some Canadian aircrew, including LMF cases, at a Canadian station in Britain beginning in the fall of 1942, but a dearth of proper facilities prompted the RCAF to establish its own central board in London in May 1944, whose staff of twelve officers included a psychiatrist and a neuropsychiatrist.[114]

These were important changes because they recognized that "fundamental differences between RCAF and RAF psychiatrists on the method of handling [LMF cases]" existed. The overall Canadian approach in LMF cases was described as "more lenient" than the RAF's. Canadian psychiatrists held that airmen who manifested signs of "fatigue, stress, or psychological illness" usually did so for physical reasons rather than as a result of their own "wilful neglect or irresponsibility." In this way the Canadian medical authorities often made a distinction between "flying stress" and LMF that the RAF did not.[115] While this was a first step toward treating Canadian airmen in a manner consistent with medical practice in their own country, the official history of the Canadian medical services in World War II has observed that since the RAF was responsible for the medical care of Canadian aircrew overseas, there was persistent dissatisfaction among "many" of them owing to differences in treatment.[116] And such dissatisfaction could be acute if an airman believed that he was being dealt with unjustly in a case of suspected LMF.

In assessing the development of aviation psychology, we can see that it underwent rapid growth because of the demands of total war on the belligerents. During World War I, when the field began to take shape, research in Britain depended on the work of a dedicated band of MOs, none of whom were specialists in the new disciplines of psychology or psychiatry. Nevertheless, a great deal of practical knowledge was acquired, and then made available in books and medical journals in the 1920s and 1930s, although little original research was carried out during the interwar years. But there was one crucial development between 1918 and 1939: the rise to prominence of civilian psychiatrists and neuropsychiatrists, who replaced the MOs with little formal training in the science of human behaviour as the RAF's official advisers on aviation psychology, and thereby became part of the British military's technocratic élite.

There is no doubt that both civilian and military psychiatrists worked very hard to ameliorate psychological disorders among Commonwealth aviators. They published numerous reports that interpreted the situation as they understood it. They realized that they did not have all the answers, and that the "pattern of war neuroses is greatly influenced by doctors."[117] Against opposition from some senior officers they promoted the contributions that psychiatry could make to the war effort, especially in questions of manpower.[118] In this way, they left their own stamp on the treatment of mental disorders among aircrew, just as the psychologists did in the matter of aircrew selection.

The psychiatrists based their treatment methods on their interpretation of some aspects of Freudian analysis, which was in widespread use during the 1930s and 1940s.[119] Today, however, scientific opinion holds that the Freudian (or developmental) and traditional medical models for treating combat stress were seriously flawed. Work done since the end of World War II has found that combat stress is "a predominantly social phenomenon." Present-day therapies focus on the soldier as a part of a larger military support system, in which such things as leadership, unit cohesion, and esprit de corps are vitally important. The reintegration of the psychologically injured combatant into his or her unit, after rest and simple medical treatment, is a priority. The emphasis has shifted from treating the patient as an isolated individual toward rebuilding the group support that allows one to endure the stressors associated with battle.[120] The assumption is that the vast majority of fighting men and women are normal people under abnormal stress, just as Birley suggested after World War I. In this analysis it is the larger military organization's role to provide the physical and psychological resources necessary to allow the individual soldier to endure the strain of combat.

With respect to theory, the greatest shortcoming of the RAF Medical Branch in its treatment of aircrew psychological disorders during World War II was the undue weight it placed on individual predisposition to breakdown. Relying on psychological theory as interpreted by its civilian psychiatrists-in-uniform, the RAF concentrated on weeding out "bad characters" instead of employing the simple treatments that had been found effective in World War I.[121] While recognizing the importance of morale, the RAF chose to focus on medical and psychiatric care for individual airmen considered salvageable, rather than concentrating resources on improving conditions that bolstered esprit de corps. Airmen deemed irretrievable were dealt with administratively. In a way this was understandable. Specialists were predisposed to use the methods they knew best. In addition, many of the potential remedies, such as improved leadership and morale, were in the hands of an executive branch that was prepared to accept that there were a few bad apples in its fighting force, but not ready to admit that its own policies might have caused unnecessary psychological casualties. And not all the remedies proposed were achievable during wartime. In most cases there was a limit to what could be done without seriously affecting operational capabilities.

From the individual airman's point of view, the most serious weakness in the RAF's treatment of psychological casualties was the unreliability of diagnosis. However, it was not discovered until after 1944 that psychiatric opinion was not dependable in individual cases, and this no doubt had serious repercussions when, in cases of suspected LMF, specialists recommended action to the executive branch.

The RCAF, without its own medical branch before 1940, and without its own medical organization overseas, had to rely on an inadequate American text on aviation psychology and on RAF advice that was incompatible with some Canadian psychiatric practices. When it was realized that British standards of treatment were different from Canadian practice, steps were taken to involve Canadian specialists more closely with the process of dealing with psychological casualties, especially in LMF cases. However, these steps came too late in the war for many Canadian airmen who were, by default, subjected to British policies and procedures.

5 What's in a Name?
The RAF and "Lack
of Moral Fibre"

The formulation and administration of policies governing aircrew who, in the judgment of British and Canadian air force authorities, "lacked moral fibre," was surrounded by controversy. According to the RAF's Inspector General in late 1942, LMF presented one of the "most difficult of all human problems" faced by the flying service.[1] For some airmen, the threat of being labelled LMF could inspire as much fear as operations against the enemy.[2] Even today the subject of LMF engenders heated debate. It remains a politically delicate issue, and the fragmentary nature of the official record concerning LMF has aroused suspicion in some, and has created what John Terraine has called "a most surprising state of affairs" in the evidence. Others have suggested a cover-up, given that anxious officials directed the RAF in late 1945 to discontinue using the term LMF so as to avoid government embarrassment at questions about the handling of airmen labelled LMF.[3] Given the unease with which the Bomber Command campaign, with its untold number of German civilian casualties and massive destruction of cities, was viewed once victory was assured, and the continuing furore over Canadian participation in the aerial destruction of Germany, we can see why historians have been circumspect about discussing such sensitive issues.

The designation LMF was employed by the British Air Ministry in World War II as a means of handling aircrew who would not or could not fly for reasons that were considered unjustifiable. The policy had far-reaching effects on air force personnel resources because its application wasted valuable manpower and was found to be unfair in some cases. In Canada the question of how to categorize airmen who were viewed as unsuitable for operations was a sensitive one, as the vast majority of Canadian flyers, whether serving in Commonwealth or British squadrons, were subject to British rules and procedures, and it was not until 1944 that the RCAF implemented its own autonomous process to deal with Canadian airmen who might be labelled LMF.

Those who seek to understand the history of the LMF question run into the problem of terminological confusion. Terraine has said that LMF was equivalent to "the old British army's 'Cowardice,'" meaning the "'inability to master fear.'"[4] John McCarthy, author of the most detailed historical analysis of LMF policy, has stated that LMF had the same meaning as the terms "Waverer" and "w case." LMF, according to McCarthy, was the name given to two kinds of behaviour that could not be dealt with under the heading of "flagrant cowardice." The first involved aircrew "who gave the impression of carrying out their duties, but who nevertheless had lost the confidence of a commanding officer," and the second involved aircrew "who openly admitted they did not intend to fly."[5] Norman Longmate has referred to those who refused to fly and to those who regularly returned without bombing the target as LMF cases. Using the expressions current among Bomber Command flyers, he has referred to some of those who had lost the confidence of their CO as "fringe merchants" (those who bombed on the edge of a target away from the defences) and "boomerangs" (those who returned early from a raid, usually shortly after take-off with a mechanical fault that could not be duplicated on the ground).[6] Max Hastings has claimed that LMF was never precisely defined and that it was applied to medical and morale problems as well as to cases of cowardice.[7] The official history of the RAF medical service rarely mentioned LMF, and, implying that LMF was not a medical problem, failed to attach a precise meaning to the term.[8]

The expression LMF predated the beginning of fixed-wing aviation and may have been rooted in late-Victorian social values, with their emphasis on "duty," "rugged individualism," and "playing the game." It was used in print as early as 1884 in the *Boy's Own Paper*, in which a character named Fanshaw was described as "a boy in whom bad instincts had been nourished by his training, and who, from constant lack of moral fibre, had gradually deteriorated."[9] In this language we can recognize the notions of "bad innate tendencies" and "poor training" that were employed by the psychological community between the wars. The *Boy's Own Paper*, however, described a predisposed individual who, lacking the necessary moral fibre in his cerebral diet, developed an inadequate personality. Later usage turned the meaning away from an item in the mental menu toward a defect in the sinews of the mind or character. It is unclear how this shift occurred; in fact, it has not yet been determined how the term LMF made its way into the RAF.[10] In any event, "LMF," "waverer," "w case," "loss of confidence," and "lacking in confidence" came to be synonymous in everyday RAF usage.

RAF medical specialists, on the other hand, preferred to avoid this terminology altogether. While LMF and waverer were sometimes used by MOs in their reports to the Air Ministry,[11] psychiatrists and neuropsychiatrists condemned such language as "unscientific." They recommended that physicians "confine [themselves] to a statement of medical matters" and not attempt to diagnose "cowardice." Two RAF MOs, R.N. Ironside and I.R.C. Batchelor, authors of the textbook *Aviation Neuro-Psychiatry* (1945), provided a rationale for absolving the medical profession of any responsibility for labelling aircrew LMF:

In wartime, loss of confidence, depending on your point of view, may be a euphemism for cowardice or an arbitrary lay concept which (*pour encourager les autres*) allows scapegoats to be selected for exemplary sacrifice from among the milder cases of neurosis. Perhaps fortunately, because scientists are unsure about the problem of responsibility, the borderlines between loss of confidence and an anxiety reaction, and between malingering and hysteria, are usually set for the doctor by the society he lives in.[12]

Such views notwithstanding, RAF doctors were involved in establishing "the borderlines ... between malingering and hysteria," and between culpable behaviour and excusable illness. In early 1940, according to John McCarthy, the Air Ministry issued its first official policy for dealing with airmen who would "'not face operational risks.'" Only those who refused to fly were to be labelled LMF and "dismissed from the service." Those who were deemed unable to carry out their duties were to be held back from promotion. At the same time, the latter were to be given "careful handling," and perhaps some leave, in an attempt to return them as quickly as possible to operations.[13]

As psychological casualties mounted in 1940 and 1941, the Air Ministry decided to create a more detailed procedure for handling LMF cases. What came to be referred to as the "LMF Memorandum," first issued in September 1941, and revised in 1943 and 1945 for greater clarity,[14] targeted, in its own words, "members of air crews who forfeit the confidence of their Commanding Officers in their determination and reliability in the face of danger in the air, owing either to their conduct or to their admission that they feel unable to face up to their duties." It placed aircrew who were "found unable to stand up to the strain of flying" into three categories. The first encompassed those who were medically fit, but who had forfeited the confidence of their COs "without having been subjected to any exceptional flying stress." The second was for those who were medically unfit "solely on account of nervous symptoms and without having been subjected to any exceptional flying stress." The third was for anyone who was medically unfit for reasons not described in the first two categories. The document emphasized that category one was only for airmen "*proved* to be lacking in moral fibre."[15] If there was any evidence of a medical disability, the airman was to be placed in category two or category three, as appropriate. The unit MO, or, on a referral, the specialist in charge of the case, was responsible for determining whether there was "evidence of physical or nervous illness." If no such evidence could be found, the case was to be dealt with by the executive branch. Despite the fact that most RAF doctors had no training in psychology, the unit MO was, in most cases, "encouraged to take the responsibility of the decision on himself."[16] The Air Ministry thus placed the burden of classification firmly on the shoulders of the medical profession. The official medical history of the RAF claimed that this occurred because the "executive were often loath to accept responsibility and preferred disposal ... through medical channels whenever possible."[17]

The administrative action prescribed by the LMF Memorandum required the airman's Air Officer Commanding (AOC) to recommend a course of action to the Air Ministry, based on the medical report and a report from the airman's CO. The final decision on how to dispose of the case was kept at the Air Ministry level, to ensure that each case received "the full consideration that it deserves both from the service and from the individual's point of view." If, before a final decision had been taken, the AOC advised that removal from flying duties was in order, the subject was to be taken from his unit and sent to a holding depot. Unfortunately, because of "executive inefficiency" some men spent months at their stations after being removed from flying duties. This was demoralizing for all concerned.[18] In the interests of fairness, the airman was entitled, at every stage of the procedure, to know the contents of any reports and to have an interview with his AOC. Beginning in February 1943 no airman could be categorized LMF if he was on his second tour of operations.

For flyers branded LMF by the Air Ministry, the consequences could be serious. Officers were required to resign their commissions. Non-commissioned aircrew were usually demoted to the lowest rank in the air force for at least three months and assigned the most menial jobs, although in some cases they could be remustered to their original trade at the rank they held before becoming aircrew. From 1944 on, any LMF case released from the air force could be called up to work in the coal mines or drafted into the army. In all cases permission to wear the flying badge was withdrawn for airmen in categories one and two, and the service documents of category one personnel were "marked in the top right-hand corner with a large red 'w'" until the practice was abandoned in October 1944.[19]

The British Air Ministry's LMF procedure has been represented as one carefully designed to weigh all the relevant considerations in what were recognized as difficult situations,[20] and there is some truth to this view. Local practices, however, greatly influenced how cases were handled. Divergences in local practices arose because executive action varied "considerably from station to station,"[21] and, as we shall see, because medical judgment was unreliable and specialists' diagnoses frequently contradicted one another. Max Hastings has recorded an extreme example of local variation, in the case of an unnamed station commander (a distinguished senior officer in the postwar RAF) who "punished by court-martial and where applicable by exemplary prison sentence" despite psychiatric advice to the contrary.[22] Perhaps the most infamous case of overzealous application of the LMF procedure involved a Bomber Command sergeant wireless operator (air gunner) who lost his leg in an aircraft accident at an OTU in June 1941. Even though he bore no responsibility for the accident, and despite his request to stay on in a ground trade, he was reduced in rank to AC2, stripped of his flying badge, and forced to accept a discharge from the air force, treatment "indistinguishable from the treatment" of a waverer. In a report to senior air force officials the RAF's Inspector General called this case "a grave scandal." He applied the same language to instances where airmen under suspicion had been forced to endure long delays before their cases were disposed of. The report also revealed that airmen in circumstances more questionable than the sergeant's had received much more lenient treatment. One junior officer, under training as a pilot, and "suffering from an anxiety neurosis," had been recommended for work as an intelligence officer.[23]

While some of the situations criticized by the Inspector General could have been attributed to bureaucratic ineptitude, others were the result of official policy. Bomber Command, as we noticed in the previous chapter, took a particularly stern view of LMF cases. Squadron Leader Reid produced a series of statistical reports for the

FPRC that supported the severe penalties meted out to many of the Bomber Command aviators who suffered mental breakdowns. One such report, based on statistics from No. 3 Group[24] compiled from August 1941 to April 1942, concluded that innate personal characteristics were responsible for the majority of breakdowns, and that the "supreme importance of the constitutional factor" dictated the early removal of those showing signs of "'Flying Stress.'" Arguing against attempts to treat such aircrew and to return them to their squadrons, Reid rejected an analogy between "the return of an infantryman to his place in the line and [the return of] a member of aircrew to flying duties," because the "consequences of the latter's failure are borne by the remainder of his crew."[25]

Group Captain Symonds, the RAF's senior consulting neuropsychiatrist, adopted a more flexible approach. While Symonds agreed that an airman "had to be removed" if stress impaired his efficiency, he stated that "exposure to flying stress is fundamentally little different from exposure to battle stress," and he noted that army statistics corresponded "nearly with our own."[26] Symonds went on to argue that there was no clear line between an anxiety neurosis and a normal emotional reaction to stress, "but that in the interests of morale, a line must always be drawn." He explained that the position of the line varied "with the group attitude towards danger," and that doctors were often called on to make an "arbitrary" decision on "whether the man had tried hard enough to satisfy group standards." Symonds concluded that "In the distinction between anxiety neurosis and cowardice expediency usually in the end counts more than scientific judgment. This should surprise no one who has reflected upon the part played by group opinion in deciding when individual behaviour should be regarded as pathological."[27] Ironside and Batchelor, similarly, recognized that "The ideal medical disposal and the ideal social disposal do not always coincide. In wartime a cleavage of theoretical opinion may be evident ... [and a] shortage of manpower and economic considerations may compel modifications of policy."[28] It is apparent, then, that influential RAF physicians in World War II, like their counterparts in 1914–18, modified their diagnoses and treatments to conform to military requirements.[29]

However, the RAF medical branch was sometimes hard-pressed to meet the demands placed on it. Owing to the rapid expansion of the RAF, experienced MOs were in short supply. Early in the war there was sometimes only one MO for each station, despite regulations that required a doctor in every squadron. Overworked station MOs were responsible for maintaining the general hygiene of their station and for the well-being of as many as 1,900 personnel. These duties restricted the amount of time they could devote to individual patients, especially those with psychological problems. And even though the MOs were expected to get to know their charges well, and to work closely with squadron and flight commanders, frequent transfers and the burden of high casualty rates often made this impossible. One MO served under twenty-two COs in a thirty-six month period, which meant he had to deal with a new squadron commander every month-and-a-half. In addition, it was especially difficult for MOs to keep an eye on the health problems of NCO air-crew, because they lived in separate messes. Even when the number of MOs increased to the point where flyers could be monitored more closely, few RAF doctors, as was mentioned earlier, had any training in psychology. Perhaps for this reason, MOs who acquired some field experience were sometimes "dubious about the value of sending flying stress cases to a neuropsychiatrist." In the end, nevertheless, most of them accepted that cooperation with specialists was necessary, particularly in treating cases outside the range of the typical MO's experience.[30]

The specialists, drawing on their own initial inquiries, believed that there was a difference between airmen who became fatigued from operational flying and those who "simply [lost] confidence." As we have seen, fatigue and constitutional predisposition were determined to be the most important reasons for flying stress. But other pressures also contributed to mental breakdowns among airmen, including uncertainties about the duration of tours of duty, lack of sleep, insufficient leave, poor physical fitness, and inadequate rest between operational tours. It was also recognized that the pressures encountered during training could contribute to the emergence of flying stress at the operational stage.[31]

In their FPRC studies, Symonds and Williams reviewed and summarized reports sent to them by RAF specialists. As in World War I, it had been found that simple therapy near the front was more effective than extended hospitalization, and most of the specialists who communicated with Symonds and Williams advised against "harsh criticism" of airmen suffering from psychological disorders, prescribing instead "tolerance and patience." They also advocated promoting "positive expectancy," that is, restoring the patient's belief that he would fly again. In the majority of cases, simple treatment involving reassurance and rest, with no suggestion to the patient that he was ill, was recommended. Mild cases of neurosis that could not be dealt with at the unit level were to be sent to the RAF Convalescent Depot Blackpool, or to the Officers Hospital Torquay, while more severe cases were to be referred to a Not Yet Diagnosed Nervous/Neuropsychiatric (NYDN) Centre (an RAF neurosis hospital) for treatment by a specialist. Some doctors, however, felt that if the problem was "constitutional or due to a faulty upbringing, psychotherapy [was] likely to be unsuccessful." In their view, the best way to handle such patients was to discharge them immediately from the flying service.[32]

Bomber Command favoured a harsh approach toward the treatment of psychological casualties. In 1940 the SMO of No. 5 Group recommended that the fate of waverers be made generally known, so that others might be deterred and the incidence of psychological disorders in the Command reduced. In a January 1941 report a Bomber Command station MO stated that "the established policy of his Group" was to rapidly remove "cases lacking moral fibre from the station." At the same time, Bomber Command's Principal Medical Officer expressed the opinion that "temperamentally unsuitable members of air crews" and "those lacking confidence ... should be given no sympathy and should be dealt with by the Executive as early as possible."[33]

By the middle of 1942, with no end in sight to the war, a series of new policies was devised to reduce psychological losses. On the recommendation of the Director General of Medical Services,[34] tours of fixed duration and regular periods of leave were introduced, and aircrew were given a rest after their first thirty "ops" (Bomber Command parlance for operational sorties). By 1943, as the shortage of experienced flyers worsened, aircrew-employment policies were modified again by extending the length of time flyers could expect to spend in Bomber Command. The Air Ministry's goal was to maintain the "efficiency and confidence" of aircrew over the course of their first operational tour, their intervening tour as instructors, and their second operational tour. If a squadron commander judged that an aviator had been "exposed to exceptional stress or [that] his stamina appear[ed] to be subnormal," he could be given special leave or early posting from operational duties without recourse to formal medical channels. Most cases were to be handled locally; however, as Symonds admitted, the proportion of cases that a unit MO was willing to treat depended on "his sagacity and enthusiasm,"[35] and the manner in which cases that were similar to one another were disposed of varied from MO to MO. Symonds acknowledged that there was real "difficulty in making the distinction between lack of confidence and neurosis."[36] Specialists required evidence "that the fear state [was] so persistent or recurrent that it [was] disabling" before a determination of anxiety neurosis could be made.[37]

But even for specialists it was no simple matter to arrive at an accurate diagnosis. Symonds eventually endorsed the findings of the

FPRC report of October 1944 on the reliability of psychiatric opinion in the RAF (discussed in chapter 4). Regarding LMF, the report concluded that in suspected cases of "lack of confidence" among aircrew, psychiatrists agreed on a determination of LMF "in only half the cases."[38] Symonds noted similar difficulties among other physicians treating flying personnel who had psychological disorders. He observed that squadron MOs emphasized "the criterion of volitional control" in their diagnoses. When "confronted with a state of fear" in a flyer, the MOs usually looked to the behaviour of other squadron members, in an effort to gauge whether the airman could be expected to "overcome" his symptoms. Pointing to his World War I experience as an infantry MO, Symonds stated that he had "long [been] convinced that the dividing line between anxiety neurosis and normal fear in combatants is artificial and related to circumstance." By way of illustration, he cited American studies undertaken in Tunisia, which showed that an infantryman with psychiatric symptoms did not always become a psychiatric casualty if he was "*not permitted* to be one."[39]

The American studies, like most of the psychological research conducted by the Allied military forces in World War II, were deficient in a crucial respect, however. Because they concentrated on the abnormal states of mind encountered by psychiatrists and by other behavioural specialists, they were concerned much more with fear than with its opposite – courage. It is only recently that researchers have begun to shift their focus from fear, especially neurotic fear, toward courage. While current research supports Bomber Command's 1944 definition of courage as "a state of mind in which fear is present but is controlled and mastered for the sake of attaining an object,"[40] it also identifies the feeling of having some control over one's environment or destiny as a key factor that permits the combatant to overcome his or her fear. While control may seem impossible to achieve in air warfare, there are certain things that can give an aviator the feeling of control. For example, if the aviator considers himself or herself a competent military professional, able to perform tasks that appear to give him or her a measure of influence over events, fear is reduced.[41]

Postwar students of fear in air warfare have closely examined the phenomenon in an attempt to reduce its effect on combat flyers. Stanley Rachman, a professor of abnormal psychology and the author of *Fear and Courage* (1978), has analyzed the seventeen volumes of aviation psychology research reports produced by the United States Army Air Forces (USAAF) after World War II, and has found that they supported many of the findings of RAF investigators. The USAAF discovered that, even though civilians of all ages and states of health could demonstrate a level of resistance to fear (under air attack for example) equal to that of combat troops, military training gave service personnel the advantage in being able to perform complex tasks in highly stressful situations. Almost every airman, the USAAF learned, experienced fear, but was able to keep that fear under control for a time. John Flanagan, the psychologist who co-ordinated the USAAF research, said the following about the flyers he studied:

The primary motivating force which more than anything else kept these men flying and fighting was that they were members of a group in which flying and fighting was the only accepted way of behaving. The air crew combat personnel were closely knit together. First, because they flew, and second, because they fought. In combat operations they lived together and had little contact with people outside the groups ... The individual identified himself very closely with the group and took great pride in membership in the group.[42]

This observation is consistent with more recent research in military psychology, which has determined that the most important social support system in combat is the group that the individual combatant works most closely with (usually referred to as the "primary group").[43] This helps to explain why the greatest fear of World War II aviators on their first missions was that of letting their comrades down and being labelled a coward. According to Rachman, once crew members had proved themselves in combat, "fears of personal failure were surpassed by fears of being killed or wounded."[44]

Based on his wartime studies of thousands of aircrew, Flanagan concluded in a 1947 report that "the seven most prominent factors in reducing fear [were] confidence in equipment, confidence in crew, confidence in leaders, continued activity, observation of a calm model [preferably the aircraft commander], [and a] circumscribed tour of duty."[45] No matter how much the stressors causing fear could be reduced individual predisposition to breakdown was still thought to be an important factor. However, postwar studies showed that there was no reliable method of identifying which individuals were most vulnerable to breakdown. While some attributes correlated well with measurable criteria (for example, flying skills with aptitude and motivation, and high intelligence with the ability to control emotions), no such correlations were established for individual predisposition to breakdown in combat.[46] A study of 150 "successful" airmen, for instance, had found "[e]ntirely contrary to expectation,"[47] that about one-half were predisposed to "emotional instability" and that nearly one-third displayed "psychoneurotic tendencies."[48] Yet all 150 "had completed a tour of severe combat duty" in which "the crews had little chance of survival."[49]

Many of Flanagan's conclusions had been anticipated by a report on the first year of USAAF combat in Europe (1942–43) prepared by the Eighth Air Force's Psychiatric Department. In the study of 150 airmen just mentioned, 95 per cent of the crews presented a "reaction to combat stress" or "symptoms of operational fatigue." Despite personality differences, moreover, the stress reactions "of the men were remarkably *alike*."[50] Even with the psychological pressures engendered by the high losses of that year, there were no courts martial of American flyers for showing fear or refusing to fly, and no cases of "true malingering." The report also said that "even the intimate knowledge [MOs] had of flying personnel in the original Squadrons and Groups that came to England did not help much in predicting" psychological casualties. MOs were "frequently ... surprised" by who did and who did not "stand up to combat." Contrary to the RAF findings discussed in the previous chapter, the report's authors discovered that the "rate of psychiatric failure [was] approximately the same in officers and enlisted men engaged in aerial combat."[51] They concluded that psychiatric prophylaxis was of little help for those airmen who broke down early in a tour, "unless there were some method of instilling an exceptional feeling of morale and the 'will to fight' in these men." For cases of operational fatigue, "rest and relaxation" were prescribed, and to this end the USAAF established, on the report's recommendation, "Rest Homes" that gave flyers the chance to spend seven to ten days in a comfortable location where the "military atmosphere" was kept to "a minimum."[52] Citing what it called the "reasonably large problem" of non-physical illnesses and "'lack of moral fiber'" among American aviators, the report also recommended that the USAAF establish "a more specific administrative policy" for cases of "Psychological Failure," one that would apply "uniformly" to the Army Air Forces.[53]

The American experience further illustrates how terminological chaos has contributed a great deal to the haze that envelops the subject of psychological disorders among aircrew. Before proceeding further it is essential to attempt to clarify some of the concepts that are useful for understanding the psychological phenomena associated with combat.[54] The term Combat Stress Reaction (CSR) is now accepted as a way of describing some psychological responses to battle. There are several possible definitions of CSR, but for our purposes the most useful is that given by Shabtai Noy, who has written that "all soldiers who negotiate evacuation with a reason other than being hit by a direct enemy projectile or explosive are CSR casualties."[55] This definition has been adopted here because it has the advantage of taking into account all possible diminutions of manpower not associated with a flyer's verifiable physical disability. In the context of the present discussion, this would include any airman who asked to be relieved of flying duties, or was prevented from flying on exhibiting symptoms of physical or mental illness for which no organic cause could be found. It would also include those who, on bombing operations, deliberately failed to press home their attack on the target, the "fringe-merchants," the "boomerangs," and those who jettisoned their bombs en route to the target to avoid the defences by gaining altitude.

The primary cause of CSR is "fear of annihilation" in battle. The contributory causes are those things that deplete one's psychological resources, such as fatigue, physical discomfort, and non-battle-related stressors, all of which can often be borne more easily when one's fighting unit is successful in the field. It has been found that CSR casualties multiply among combatants on the losing side, and when the number of wounded in a unit increases significantly.[56] Investigations conducted during and since World War II have found that unit leadership and cohesion are the primary supports that keep combatants going in battle. Once the social support network provided by leaders and by the unit disintegrates, people feel overwhelmed and CSR casualties mount dramatically. Research has not been able to identify any predisposing traits to CSR; however, some people may have attributes that make them more resistant to stress.[57]

Modern treatment of CSR is similar to that discovered to be effective in World War I, especially the strategy of not labelling CSR casualties as "sick." Helping those suffering from CSR to maintain an image of themselves as essentially healthy people who, with some rest, can successfully cope with their condition, has been found to be important in aiding quick recovery.[58] Interestingly, this approach bears some resemblance to the treatment model used at times in World War I, in which aviators were encouraged to think of themselves as tired and in need of a rest, rather than ill.[59]

The current psychological literature suggests that "strength of leadership and unit cohesion are the only factors with demonstrated merit in reducing [CSR] casualties." Strong leaders, in this context, are those who take prophylactic measures to reduce the incidence of CSR in their units. There are five such measures in vogue today. First of all, the experts recommend training that promotes self-confidence in military personnel and familiarizes them with the tasks that will contribute toward unit success in battle and take their minds off the stresses of war. Second, keeping military personnel informed about the course of battle operations has been shown to enhance their coping resources; commanders are counselled to keep those who serve under them as fully informed as possible, even if the news is bad. Third, commanders are encouraged to be aware of and respect the principles of csr treatment in order to reduce psychological casualties in their units. Fourth, commanders are urged to work closely with unit medical personnel to establish personnel-conservation strategies, and to devote unit resources to this goal to the fullest extent possible. This includes making the mo a fully participating member of the unit, and allowing him or her to share in its aspirations and concerns. Finally, commanders are advised to do everything in their power to keep their people together as a group, and thus to reap the psychological benefits of unit cohesion.[60]

Virtually every commentator on bomber operations in World War II supported principles similar to these and concluded that successful units led by trusted commanders had fewer stress-related casualties.[61] Donald Schurman, a Bomber Command veteran interviewed by the author, remarked that his fellow aviators were motivated by "the glucose pill of ops." The five ingredients of this "pill" were, first, pride in one's own performance and in the performance of one's crew, squadron, and group; second, faith in one's leaders, at all levels; third, confidence that the skill of one's crew would enable one to survive an operational tour; fourth, a belief that one was taking part in successful operations against the enemy; and finally, a conviction that the war against Hitler was just.

Recent research confirms what was learned about leadership in World War II, namely that the most trusted commanders are those who demonstrate professional competence; are credible in their role as purveyors of information; and devote care and attention to their subordinates.[62] For air forces, we should add another characteristic: the willingness of commanders to share their subordinates' risks from time to time. In World War II, Leonard Cheshire epitomized these qualities, and the record of his command demonstrates the importance of leadership in the csr equation. Cheshire, whose story has been told by Max Hastings, was co of Bomber Command's 76 Squadron during the first four months of 1943, and was one of the raf's most celebrated leaders. Some cos got the derisive nickname "François" from their subordinates because they usually participated only in relatively safe raids on France, but not Cheshire. He deliberately elected to fly "with the new and the nervous" as second pilot on dangerous raids. In this way he demonstrated his competence and his willingness to take risks. He also "inspired loyalty and respect" because he made a point of getting to know all of his air gunners (the aircrew with the lowest status in the eyes of many in Bomber Command) and ground crew. By the end of the war Cheshire had earned a Victoria Cross, three Distinguished Service Orders, and a Distinguished Flying Cross, and had become "a legend."

His replacement had a much different experience. Rarely flying on dangerous ops, and plagued with flyers who returned early from their missions, claiming "bad luck," the new CO saw the unit's LMF rate increase alarmingly. By late spring 1943, 76 Squadron's early return rate sometimes exceeded 25 per cent of the aircraft dispatched. Not surprisingly, at the end of 1943 this CO was replaced. His successor, "Hank" Iveson, resumed the custom of the CO flying on dangerous missions, and also broke up crews with a history of early returns. The improved unit performance significantly ameliorated morale, but to maintain morale at a high level the CO had to be constantly on the alert. When the squadron was re-equipped with the new Mark III Halifax, which had a "fearsome reputation for accidents," Iveson and his three flight commanders flew on the first mission to demonstrate their confidence in the aircraft. Throughout his tour, Iveson elicited excellent results from his crews.[63]

The example of 76 Squadron demonstrates what was achievable by charismatic leaders. It also proves that the men of Bomber Command could not be driven to their tasks; there were too many ways for them to shirk their duty, especially on night operations, if they felt their leaders were letting them down.[64] But no matter how inspirational the leadership, there was a limit to what anyone could endure. Bomber Command had recognized this fact after its first few months of operations, and in February 1940 headquarters suggested that some provisions be made for the "regular relief of war-weary air crew," although no structured scheme was introduced.[65] Leave facilities were provided to which aircrew could be sent at the discretion of group commanders and group SMOs. This measure was intended to conserve manpower by saving some flyers who needed a rest, but who under previous procedures would have immediately been struck off unit strength.[66] As one Bomber Command SMO remarked about this period in its history, "Flying personnel used to say that they flew till it was 'coffin or crackers.'"[67] In other words, they flew, without hope of survival, until they were killed or went mad. In March 1941 Bomber Command introduced the fixed "operational tour," which was designed to allow aircrews to rest after a set period in action, and to inspire them with the knowledge that they would not have to endure combat indefinitely. However, training-staff shortages caused almost all who finished their tour to be sent immediately to OTUs,[68] where the great numbers of accidents and the psychological stresses of flying training gave rise to a casualty rate similar to that of combat. As one eyewitness put it, "flying is one of the few activities where training and preparation exact a casualty rate comparable with combat."[69]

Bomber Command senior staff acknowledged that, under these circumstances, everyone "had the windup" but also said that proper training, effective discipline, high morale, confidence in equipment, and good leadership would allow most aircrew to overcome their fears. Operational conditions, especially bad flying weather and improved enemy defences, also contributed to stress. However, squadron commanders believed that some circumstances were controllable, and they specifically underscored the disastrous effects on morale of repeated cancellations of missions, especially last-minute cancellations. With respect to morale they felt it was better to attempt a sortie in bad weather than to cancel late. One critic of mission cancellations cited the case of a "freshman" who "scrubbed" seventeen times before his first op; when he finally started to fly operationally, he quit after three trips. This pilot had endured as much stress before he got airborne against a real target as had someone who had been sent out on many ops.[70] The problem of continual cancellations was particularly bad during spells of poor weather, especially in winter. Periods of low operational activity were observed to have a greater negative impact on morale than the higher casualty rates that accompanied periods of intense activity. Among the things that kept the men going, "crew spirit" was acknowledged to be essential.[71] Some senior officers commented that it was important for aircraft captains to be good pilots and good leaders, but that unfortunately it was "difficult to find both [qualities] in the same man." In the eyes of some commentators, the best way to avoid combat stress was to give the men proper training. This would build confidence, and lead in turn to "fearlessness." Badly trained men were observed to suffer more from fatigue than those who were well-trained.[72] As for the role of discipline, there was a noticeable difference of opinion among Bomber Command leaders. All agreed on the necessity for air discipline, that is, strict adherence to flight procedures, but there was no consensus on whether an emphasis on discipline on the ground aided or hindered discipline in the air.[73]

But good leaders needed to be more than disciplinarians. The RAF was well aware of the importance of leadership and esprit de corps in promoting unit efficiency and "in preventing or causing psychological disorder in flying personnel."[74] Even a service psychiatrist was moved to declare that "'What these people really need is not a psychotherapist, but an evangelist.'"[75] Evangelists, however, were in short supply.

One month after the war began, there was concern in Bomber Command about the abilities of its commanders, especially at the squadron level. Charles Portal, the force's c-in-c at the time, stated that squadron commanders in their forties with minimum flying hours were "utterly useless," and that he needed competent pilots aged twenty-six to thirty-five to replace them. Similarly, he wanted station commanders who were "active and alert," and preferably under age forty.[76]

In mid-1941 a No. 6 (Training) Group station commander, Ruscombe Smyth-Pigott, recommended that all group and station commanders fly every four to six weeks with their crews, especially in bad weather. He said that the example of senior officers flying on ops would be good for morale, and they would also see firsthand the results of their tactical decisions. Senior officers not keen to do this should, he declared, be "swept aside."[77] An example of the type of leadership favoured by Smyth-Pigott was in evidence during Bomber Command's first thousand-plane raid, which took place on the night of 30 and 31 May 1942. The risks were high, and Churchill was prepared to lose 100 aircraft.[78] In the face of this danger one station commander made the following announcement at the crew briefing:

The c-in-c says you will spread apprehension and despair throughout Germany ... I have therefore delegated my duty in the Ops Room ... in order to satisfy my pleasure in observing your firework display from the rear turret of "A" Flight Commander's aircraft.[79]

By choosing to fly on perhaps the most dangerous raid yet carried out by Bomber Command, and in the most hazardous position in the aircraft, this station commander served as an inspiration to his men, but he was not the only high-ranking officer to join the thousand-plane raid. At least one group commander and "several" station commanders also flew "with their men" on this raid.[80]

These examples suggest that Bomber Command solved its leadership problem to its satisfaction, at the station commander level and above, by early 1944. By then, according to Arthur Harris, these senior officers each had an average of about fifteen ops to their credit, and were, therefore, active and aware of the tactical situation.[81] Besides these actions, some other solutions to the leadership dilemma in Bomber Command came from inquiries conducted by the medical branch.

A good deal of light was shed on the question of leadership in Bomber Command in a study conducted by Symonds and Williams in August 1942. After interviewing forty-four aircrew, mostly station, squadron, and flight commanders, and thirty-seven MOs, they concluded that good leadership was "vital' to helping men "accept and carry the load of operational flying." While there did not appear to be any one type of personality that ensured good leadership,[82] there were a number of personality characteristics and behaviours that were generally recognized as the hallmarks of successful leaders. The first task of a good leader who was new to a squadron was to establish his expertise in flying. If he had had no operational experience in Bomber Command, he needed to demonstrate as soon as possible that he was "an efficient operational pilot." It was also important that he share in the risks faced by his squadron by going on "difficult raids," especially "when losses [were] heavy or morale low." On ops, aircrews wanted their CO to set an example of steadiness under pressure. They also appreciated a keen commander who displayed initiative and drive, and wanted to believe that a CO's "whole interest" was in the squadron. This would be demonstrated by his acquiring "a personal knowledge of all the crews," and by being accessible to them when required. At the same time, Bomber Command aviators expected a CO to be hard but fair "in all matters" connected with operations. Above all, he had to be seen as a leader; it was recognized that even a "very good pilot may be a bad leader." To foster a perception of good leadership a CO was expected, particularly "when things [were] going badly," or after "very heavy casualties," to take an active interest in his squadron by organizing "intensive training." Speaking "quietly and with confidence," and going on an op when the squadron was having a run of bad luck, further built up the image of a concerned, effective leader.[83]

The importance of leadership was such, according to Symonds and Williams, that "the fortunes of the squadron" were often described "in terms of its commanders." One station commander remarked to them that cases of lack of confidence "usually occur[ed] in epidemics, and when an epidemic occurs it is usually due to a bad squadron or flight commander." In one instance, when "it became known that a [new] squadron commander wouldn't fly operationally," five cases of LMF occurred within the first fortnight of his arrival. Men cracked "because they had no confidence" in their leaders.[84] It is in one sense ironic that LMF cases proliferated when COs lost the confidence of their men, for this is precisely the opposite of what the LMF memorandum envisioned.

At this point in the discussion it is necessary to turn to the much neglected issue of the number of psychological casualties among aircrew. According to Terraine, "the only available statistical analysis" is Wing Commander J. Lawson's study of 2,726 cases classified LMF. The number labelled LMF from July 1943 to June 1944 represented less than 0.3 per cent of all RAF aviators and 0.4 per cent of Bomber Command aircrew.[85] These figures could lead us to assume that Bomber Command was almost immune to losses from psychological causes,[86] but that was not the case. As we have seen, the lack of an agreed-on medical definition of "flying stress," and the vagaries of administrative action on LMF cases, make Lawson's statistics virtually meaningless, for they include, for the most part, only those aircrew who were unfortunate enough to be given a label – LMF – that many others avoided.

A much more useful exercise is to examine the number of psychological casualties suffered by the RAF, and by Bomber Command in particular, rather than focusing solely on LMF cases. Yet this is not an easy task. There was no uniformity in the way statistics were collected, for one thing. For another, the processing of psychological cases was the product of a multiplicity of decisions, executive and medical. The result was a welter of conflicting and confusing data. There is, however, something to be gained by examining the records kept by the RAF Medical Branch. For the purposes of the present analysis other data sets that might bear on the question of psychological disorders among aircrew, such as the records pertaining to flyers who returned early from their missions without justification, will be excluded. Owing to the difficulties associated with the data, the following discussion will establish orders of magnitude rather than precise figures. It is, however, a beginning in the exploration of heretofore uncharted territory.

The most complete and consistent analysis of psychological disorders among RAF aircrew was undertaken by Symonds and Williams, and from it we can obtain some idea of the incidence of psychological casualties in Bomber Command. During the three years from 10 February 1942 to 9 February 1945, according to an FPRC report prepared by Symonds and Williams, 8,402 RAF aircrew were diagnosed by the service's neuropsychiatrists as suffering from a "neurosis."[87] To this figure we should add the 1,029 found by the same specialists to be "lacking confidence," the 108 "not having a neurosis but unsuitable for flying duties," and the 129 "considered normal and returned to duty," for as Symonds and Williams themselves admitted, the aircrew in these three categories were probably neurotics whose mental state had been misdiagnosed, and in any event their manpower was lost to the service.[88] In total, then, 9,668

RAF aircrew were removed from flying duties by the service's neuro-psychiatrists during the three years for which such statistics exist.[89]

Beyond this point the data are less exact, and it is sometimes necessary to use results based on small sample sizes, or to draw inferences from sources other than the medical records. Symonds and Williams found in one study that 70 per cent of psychological disorders among aircrew were diagnosed by unit MOs;[90] specialists probably saw the other 30 per cent.[91] Therefore, if the total of 9,668 given above represents only 30 per cent of the aircrew recognized by the RAF Medical Branch as suffering from some sort of psychological disorder or labelled LMF, there may have been 30,000 or more RAF aircrew so-classified from February 1942 to February 1945, or approximately 10,000 a year.[92]

Statistics gathered during he war indicated that Bomber Command (including the OTUs) accounted for about 40 per cent of all "neurosis" cases among RAF aircrew,[93] which means that of the estimated 10,000 RAF LMF and psychological cases every year, Bomber Command's share would have been roughly 4,000. Given that about 80,000 men served as Bomber Command aircrew during a typical year,[94] it follows that about 5 per cent experienced a psychological disorder that was recorded by the medical authorities. This is the same annual proportion, it will be recalled, that was found to apply to Bomber Command "operational air crews [who] developed psychological illness of sufficient severity to lead to admission to hospital or disposal by CME" in the period up to January 1941.[95] Of course, many cases were not dealt with formally by the medical system, and there is no way of estimating the proportion that escaped official medical notice. However, if we accept that approximately the same number of cases were handled informally by MOs and by the executive branch as were handled through regular administrative channels (and therefore did not show up on sick-report records), it is possible that roughly 10 per cent of Bomber Command aircrew suffered some form of psychological complaint.[96]

The exact number of Bomber Command flyers whose lives were lost as a result of a misfortune attributable to psychological causes will never be known, though we can try to arrive at an estimate.[97] An average of 20,000 Bomber Command aircrew were killed or wounded each year, the vast majority in situations where the causes were unknown.[98] However, based on the accounts of survivors and modern studies which report that even in peacetime 50 per cent of "successful combat fliers" show signs of "emotional instability," and 25 per cent of all accidents attributed to aircrew error in an acute emergency have "chronic stress" as a contributing factor,[99] it may not be unreasonable to speculate that 50 per cent, or 10,000, of Bomber Command's annual casualties were caused by airmen whose psychological state degraded their operational performance enough to lead to death or serious injury. It is not intended to suggest here that all 10,000 suffered from a "psychological disorder," but in Bomber Command it took an error on the part of just one crew member out of several to make everyone a casualty.[100] We can say, then, that about 10,000 of the 80,000 airmen in the force during a typical year, or 13 per cent, were killed or wounded because they, or one of their companions, were mentally unsound. And if we add the 13 per cent to the 10 per cent identified in the previous paragraph, it may be that psychological disorders among Bomber Command airmen caused the force to lose 23 per cent of its flyers every year.

However, the figure of 23 per cent may have been reduced by other factors. According to wartime reports, between one-quarter and one-half of the airmen treated by unit MOs or by neuropsychiatrists were returned to "full flying."[101] However, full flying was not defined, and there is no indication that it always meant a return to full operational duties in Bomber Command. Moreover, any treatment regime that removed a Bomber Command airman from operational flying for more than a few days disrupted his crew, and necessitated his replacement by another trained aviator. In the executive branch's view, the goal of psychological treatment was to produce rehabilitated airmen who were "fit and keen for a second tour," not to force them back into service in time to complete their first tour.[102] So, even in cases of successful treatment, there could be substantial delays in getting aviators back to full effectiveness. Nevertheless, if we assume, based on the above factors, that one-quarter of the 4,000 airmen who were officially diagnosed every year by Bomber Command doctors as suffering from a psychological condition were returned to full flying duties with minimal effect on operations, that leaves 3,000 who were lost to the service. But as we saw earlier, we need to double this figure to account for those who were dealt with outside formal channels. Add the doubled figure, 6,000, to the 10,000 deaths mentioned above, and we get 16,000 of 80,000 Bomber Command airmen (20 per cent) who died or were otherwise lost to the service for reasons having to do with their own or a fellow airman's impaired psychological state.

To summarize, it has been suggested here that about one-fifth of Bomber Command's wastage can be attributed to psychological factors. It should come as no surprise that the casualty rate from psychological causes may have been of this order of magnitude. Aircrews, especially when engaged in night bombing operations, were subjected to high levels of stress, as was evidenced by the fact that the incidence of neurosis was four times higher in this role than in the RAF as a whole.[103] If the 20 per cent figure seems high, perhaps a comparison with the Canadian Army's experience may put it in perspective. It was discovered in World War II that certain infantry combat units suffered CSR-related losses of over 30 per cent, and Symonds, as we have seen, remarked that army statistics corresponded "nearly with our own."[104] The Canadian Army used the term "neuropsychiatric (NP) ratio" to describe the proportion of neuropsychiatric casualties to total battle casualties (killed and wounded). The NP ratio for Canadian infantry units engaged in battle in Italy from late 1943 to mid-1944 averaged about 23 per cent; post–D-Day fighting yielded NP ratios approximating 25 per cent among Commonwealth and American infantry.[105] In Bomber Command, as we observed a moment ago, there were about 4,000 officially recorded LMF and psychological casualties a year, out of an average annual casualty total of roughly 20,000. Bomber Command's NP ratio was thus bout 20 per cent, which comes close to the 23 per cent for the Canadian Army in Italy.

From the standpoint of total losses (psychological and non-psychological), Bomber Command suffered grievously compared to other units on what has been called the "cutting edge of battle." Canadian rifle companies in the early Italian campaigns, and British and American infantry units in Normandy, experienced annual casualty rates of 50 per cent, 76 per cent, and 100 per cent of unit strength, respectively.[106] Considering that Bomber Command's operational aircrew strength was about 8,000 in 1943,[107] and that the force suffered about 20,000 casualties annually, Bomber Command's rate was 250 per cent of unit strength. Seen this way, the incidence of psychological casualties in Bomber Command appears quite acceptable in relation to the very heavy casualties it suffered.

This analysis has focused on those airmen who suffered death, injury, or removal from flying duties because of the stresses of operational flying. To account for everyone who might have been affected by CSR, it would be necessary also to include the airmen who escaped death, injury, or removal from flying duties but whose operational performance fell short of the required standard because they were suffering from a battle-related psychological condition. We would have to include, for instance, the flyers who released their bombs early because they feared the enemy's defences. But such an exercise is beyond the scope of this study.

It should be noted, finally, that although the concept of LMF as it was employed in the RAF and the RCAF is difficult to analyze statistically, it had a rationale. It has been suggested that since all aircrew were volunteers, there was no need for procedures that stigmatized them as LMF or humiliated them in some other fashion.[108] As this chapter has shown, however, even volunteers, especially when they are poorly led, can behave in ways that subvert organizational objectives. The LMF label did inspire fear, and it did keep some flyers at their stations. Some veterans of Bomber Command have said that many of their fellow airmen approved of the LMF label as a means of discipline, for its use was necessary to keep up the bombing effort.[109] Before rejecting LMF policies out of hand we should consider the opinion of their defenders that an LMF classification was less harsh than a court martial, and an appropriate way of dealing with those who would not or could not face the strain of ops.[110] Because of the difficulties in holding courts martial on operational bomber stations, there also were a number of administrative advantages to the LMF procedure. The problems of keeping aircrew at their base for the trial, which would have had "'a most serious effect on the morale of others,'" and "'the difficulty of calling witnesses on account of casualties and postings,'" have been cited as two important reasons for avoiding formal legal proceedings.[111]

The LMF procedure, however, was most open to criticism because its rules were not applied uniformly, a situation understood by the British Air Ministry but one which it could not fully control. It was this controversy that led the RCAF to develop its own procedures for dealing with LMF cases, as we shall see in the next chapter.

6 Conserving the Cream of the Crop: The RCAF and "Lack of Moral Fibre"

In addition to the uneven application of regulations and the genuine difficulty of distinguishing most waverers from psychological casualties, the RAF's LMF procedures created other serious problems. Many of these problems were attributable to the way in which LMF policies were developed. The first rules governing waverers, as we have seen, were introduced in 1940 by the Air Ministry as a stopgap measure to deal with "acute neurosis," particularly among Bomber Command aircrew after the force had sustained heavy casualties. The rules were meant to regulate all cases of removal from flying duties where the airman's mental state was at issue.[1] As time went on, the rapid increase in the size of the combat forces under RAF command led to greater numbers of aircrew who were not fit to perform their assigned tasks. The LMF procedure was modified to cope with this problem, as was noted in the last chapter, but the procedure proved too blunt an instrument to deal with it effectively, since waverers were just one part of the larger issue of aircrew who were unable or unwilling to carry out their operational duties. As Canadian officials became more conversant with RAF policies governing aircrew removed from flying duties, they began to have serious reservations about the British Air Ministry's ability to handle RCAF airmen in a manner politically acceptable to the Dominion. During the negotiations preceding the creation of the BCATP, Canadian politicians had stressed that the Canadian people wanted to be represented by their own front-line squadrons as quickly as possible.[2] Until then, they expected their airmen serving with the RAF to be treated fairly.

One of the most sensitive issues became how to deal with aircrew who were not fit to fly operationally, a subject on which the authorities of the two countries did not see eye to eye. Although the development and application of aircrew-employment policies suffered from bureaucratic friction and administrative incompetence, the source of the controversy lay in fundamentally different approaches to the handling of aircrew suspended from flying. The Canadian authorities emphasized the salvaging of aircrew under suspension, while the British were more concerned with the maintenance of morale by using disciplinary measures to weed out those believed to be unfit. It has been suggested by several authors that the RCAF shared the RAF's unduly harsh attitude toward LMF cases,[3] but the following discussion will show that the RCAF's approach was different. By creating its own LMF and other aircrew-employment policies and the administrative machinery to handle those matters, the RCAF eventually asserted its independence in an area essential to the conservation of skilled military manpower.

During the war, Commonwealth aircrew were removed from flying duties for four principal reasons: medical conditions (physical or psychological), LMF (as defined in the LMF Memorandum), "misconduct," and "technical unsuitability" or "inefficiency." Misconduct was the term reserved for those who were dealt with under the military justice system, whether by summary trial or by court martial, for breaches of air discipline and for crimes committed on the ground. Technical unsuitability was used to describe servicemen who failed their aircrew training, while inefficiency referred to those who completed their training but were unable to perform to operational standards.[4] Early in the war, however, little distinction was made between these categories, and many who did not come under the terms of the LMF Memorandum were, nevertheless, treated as waverers.[5]

Senior Canadian officials first became intimately involved with the LMF issue during a visit to Britain in June and July 1941 by a delegation of senior RCAF officers led by Charles "Chubby" Power, minister of national defence for air (MNDA). In Power's eyes, a crucial issue for discussion was the safeguarding of "the interests and well-being of RCAF personnel whether in the Canadian Squadrons or attached to RAF units,"[6] including those who were labelled waverers by the RAF. The British Air Ministry officials who met with Power explained that an officer found to be LMF had his commission terminated and was refused employment "in a ground job in the Service." Steps also were taken "to prevent his getting a lucrative job as a pilot in civil life." NCO aircrew were reduced to AC2, the lowest air force rank, and, before they could remuster to another trade, were usually assigned to menial jobs for at least three months. RAF officials added that there was "another category of Waverers known as 'Medical w's' whose lack of moral fibre may be attributed to family background." It was claimed that they were "more sympathetically treated" and "invalided out of the Service." However, in all waverer cases "wings [were] taken away." A third group, not classified LMF but covered by the LMF Memorandum, comprised those "[c]ases which have a history of crashes, injuries, great stress, etc." They were "dealt with by transfer to ground duties," but retained their wings.[7]

The British officials who met with Power requested that the RCAF consider treating all its aircrew, particularly those not subject to British regulations, "on the same lines."[8] However, the "RCAF made no commitments ... but undertook to study" the waverer question, which Power and his advisers claimed was "not so acute in Canada."[9] The RCAF, in fact, had no regulations governing LMF cases, and none were to be made until after Power's return to Canada. In the meantime, RCAF aircrew overseas were dealt with through British channels.

In October 1941, three months after his return home, Power authorized a set of "Regulations Respecting Officers and Airmen Removed from Aircrew Duties," which established an LMF procedure very similar to the one provided for in the RAF's LMF Memorandum. The main differences were that any deprivation of flying badge or reduction in rank was only to be by order of the MNDA, and flyers who overcame "their lapse of courage" could have their wings and rank restored.[10] These regulations applied only to personnel "serving in Canada"; RCAF servicemen overseas were still subject to British regulations.[11]

However, it soon became apparent that Canadian airmen in Britain were not being treated as fairly as the Canadian public thought they should be,[12] which angered the many Canadians at home who were not prepared to have the reputations of their sons impugned by the disciplinary procedures of a foreign force. The Canadian public demanded that Ottawa take action, which ultimately resulted in Power, who also held the position of associate minister of national defence, becoming closely involved in the controversy, for a number of uniquely Canadian reasons. First, Power's superior was the politically astute minister of national defence (MND), Colonel J.L. Ralston, who took a personal interest in any detail, rumour, or criticism that could be a source of embarrassment to the government. Power knew that the Canadian public, unlike the British public, could be demanding, sometimes unreasonably so, of a Cabinet minister, for example expecting him to know every detail in the files of individual airmen. A self-described servant of the people, Power accepted this burden, and felt "obliged to give more attention to these matters than was necessary in either Britain or the United States."[13]

In World War I Power had served under British command, and had often chafed at the constraints of imperial control. He had been employed in various roles, rising from batman and kitchen helper in Britain to infantry major on the western front. Like many of his generation, he supported Canada's efforts to lessen British imperial influence, and he looked to Sam Hughes's penchant for confronting Lord Kitchener and the War Office as his model when standing up for Canadian rights against the British Air Ministry. In addition, as someone who had served both as a private soldier and as a senior officer, and had spent time in detention for being absent without leave, Power felt that he could sympathize with most airmen.[14]

Just as the government of World War I had responded to public concern for the welfare of soldiers by instituting various measures to enhance the status of the Canadian land forces, which measures in turn helped to promote national identity and autonomy, in World War II public concern for the welfare of airmen was expressed in the government policy of "Canadianization." Designed to give the RCAF greater freedom from British control, Canadianization included, among other things, the creation of Canadian squadrons and higher formations, and a transfer to the RCAF of jurisdiction over personnel policy and administration with regard to its airmen overseas.[15] Despite official British acceptance of Canadianization, Ottawa's policy was obstructed at many levels in the RAF. At higher levels it was perceived as the unwarranted political interference of a junior ally in operational matters where Canada had little influence. Consequently, some senior RAF commanders simply circumvented Canadianization when it suited them.[16] At lower levels the policy was seen as disruptive to morale, and efforts were made, especially with RCAF NCO aircrew, "to blot out" their Canadian identity by insisting that they belonged to the RAF once they left Canada.[17]

Power, the minister responsible for implementing RCAF Canadianization policies, also faced resistance from some important anglophile RCAF officers. Air Commodore L.F. Stevenson, for instance, was an outspoken critic of Canadianization. Because of his opposition the government eventually removed Stevenson from his position as commander of RCAF Overseas Headquarters in Britain,[18] and sent as his replacement a staunch supporter of the government's policy, Air Vice-Marshal Harold "Gus" Edwards, who took up Power's cudgels shortly after arriving in London[19] in November 1941. Edwards began by criticizing the way in which certain RAF policies were applied to

Canadians. Citing the need for more Canadian doctors to monitor the removal of RCAF aircrew from flying duties, he declared that the British MO did not understand Canadians. The British MO, Edwards said, "treats the malingerer sweetly and does not recognize the truly sick man." Edwards's assessment was reminiscent of the VAD controversy of World War I, when the Canadian public was concerned that its wounded servicemen "dispersed throughout" Britain were not being given proper medical care by British doctors.[20] Edwards also found that RCAF aircrew labelled LMF were being deprived of their rank and flying badge by "others than our own people," often in an abrupt and degrading manner. He declared this to be "objectionable"[21] and directed his staff to raise the matter with the British Air Ministry.

Shortly after Edwards issued his directive, Wing Commander J.L. Jackson, Edwards's Director of Personnel, informed the RAF that "There has for some time been considerable evidence of discontent amongst RCAF personnel who have been stripped of their aircrew badge for various reasons, in some instances for medical reasons. In some cases S[ergeant] aircrew have been retained at their OTUs employed as A.C.2 General Duties." He suggested that a conference be called so that the "difficulties which have beset us in the past due to policy not being completely clarified, might be removed." He also requested that all RCAF cases involving loss of flying badge *other than Waverers*" be referred to RCAF Overseas Headquarters.[22] According to Jackson, referral was provided for under "the general policy regarding RCAF personnel,"[23] but it is not clear where this "general policy" originated, as Canadians overseas who were removed from aircrew duties were, at this time, subject to RAF regulations.[24] For the moment, however, the main issue that attracted the concern of the Canadian authorities was the status of prospective aircrew who had failed their OTU courses for reasons beyond their control. The authorities believed that in most cases such men should retain their wings and rank, and be employed usefully if possible, whereas British policy required that they lose their wings and rank, and often that they perform menial jobs or be dismissed from the service.[25]

The Canadian request for a conference triggered a convoluted series of negotiations between Britain and the Dominions, negotiations that bordered on the bizarre. At the first meeting, held on 30 December 1941 and chaired by the RAF Director of Personal Services, representatives of the RCAF, the Royal Australian Air Force (RAAF), and the Royal New Zealand Air Force met with the senior Air Ministry officials charged with administering aircrew-disposal policies, including policies under the LMF Memorandum. The meeting began with a discussion of waverers, during which the delegates agreed that the LMF Memorandum should apply to Dominion servicemen. A difference of opinion then arose over how waverers should be treated. In an apparent contradiction of the letter sent to the Air Ministry by Wing Commander Jackson, the senior Canadian representative, Group Captain W.A. Curtis, Edwards's deputy, stated "that authority for the removal of badges and reduction in rank was now vested in the [MNDA]." Therefore, the details of all disciplinary cases involving RCAF personnel, waverers included, had to be referred to RCAF Overseas Headquarters for transmission to Canada before any action could be taken. The delegates agreed that "all Dominion personnel alleged to be lacking in moral fibre should be submitted to the Air Ministry under the arrangements laid down in" the LMF Memorandum. The Air Ministry officer handling waverer cases, Wing Commander J. Lawson, would then refer "proved cases" to Dominion authorities "for formal approval to proceed with normal disposal action." To preserve aircrew morale, the British delegates insisted that if an RAF "recommendation to classify [a Dominion airman] as 'w'" was not accepted by the airman's government, he would be repatriated on the grounds that "he was no longer employable in an RAF squadron."

With the LMF procedure for Dominion aviators settled, the Dominion representatives raised what they considered to be a more contentious issue. Citing "many instances ... which have caused considerable embarrassment," they restated their opposition to taking away the flying badges of their air force personnel for any reason other than "proven loss of moral fibre." At Dominion insistence, it was agreed to amend the Air Council[26] regulations governing the removal of badges so that Dominion authorities had to approve removal before it could be effected.[27] Air Marshal Philip Babington, the RAF's AMP, refused, however, to change the regulations, and another meeting was convened on 5 February 1942 to address his objections. Canada's representative, W.A. Curtis (now an air commodore), bluntly stated that "Canada would *not* agree to give the badge and then take it away"; to do so, except in the case of waverers, "would be a crime." The meeting could not resolve the issue, and so it was referred back to the AMP and the RAF Air Council.[28] In early March the Air Council reluctantly agreed with the Dominion position that removal of the badge "should be restricted to cases of proved lack of moral fibre," and RAF Commands and Groups were advised that same month that they were not to take away the badges of Dominion servicemen who failed OTU training.[29]

These negotiations also resulted in changes to the Canadian LMF procedure, changes that offered some added protection for RCAF personnel. The first stage of the Canadian procedure was identical to the RAF's. RCAF flyers defined as waverers under the LMF Memorandum were to be removed from aircrew duties and posted immediately to the RAF Reception Depot at Uxbridge for further processing. Under the new procedure, however, "the full case history [was to be] referred by the Air Ministry to" RCAF Overseas Headquarters, where it would be reviewed by the Director of Personnel, who could convene a board of officers in doubtful cases. If the Canadian authorities concurred with the Air Ministry's decision, the decision would be referred to the MNDA "for final ratification."[30] Edwards later arranged that badge removals and reductions in rank were to be "done in as nice a manner as possible."[31] NCO aircrew would have their badges removed, "[f]or the time being," by an RAF wing commander at Uxbridge.[32] They would be reduced in rank to AC2, and then "placed at the disposal of the RAF for service in any theatre of war." Officers, on the other hand, would only be reduced in rank on returning to Canada, although their badge would be taken away before they left Britain.[33]

However, there were a number of problems with the new process, the most serious being administrative delays that unduly prolonged the time suspected waverers had to wait for medical boards and documentation. While they waited, aircrew under suspension remained at their stations and, in the opinion of both British and Canadian commanding officers, their presence had an "unhealthy effect on morale."[34]

In addition, according to Air Marshal Babington, there was still "considerable misunderstanding" on the part of the Dominion representatives about the question of badge removals. And in one respect, at least, he was right. The Dominion representatives had never been sent complete copies of the LMF Memorandum or of the Air Council regulations that amplified the rules governing removal. Air Ministry officials justified this by saying that the regulations were confidential and exclusively for Air Ministry use, and that not even senior RAF officers in the field were informed of their complete contents. Faced with this complication, and with continued pressure from the Canadians to assert their own authority in LMF policy, Babington decided that the LMF question should be taken up at the ministerial level.[35] As a result, the question was referred to the next meeting of the responsible ministers.

The Ottawa Air Training Conference, which was held from 19 May to 3 June 1942, largely at Canada's insistence, to resolve a number of pressing issues related to the BCATP, gave British and Canadian government ministers a chance to review LMF policy. Britain's senior delegate to the conference was its undersecretary of state for air, Harold Balfour, whom Power admired and with whom Power maintained "cordial personal relations, notwithstanding wide diversions of opinion."[36] In return, Balfour found Power to be "grand to negotiate with" – when he was sober.[37] A seasoned pilot with World War I combat experience, Balfour understood the psychological problems associated with air warfare. Perhaps because of this, he eventually became the minister responsible for reviewing RAF LMF cases that "could involve some penalty."[38] He was asked to do this because some British parliamentarians were concerned that senior officers in the Air Ministry were "old-world 'blimps'" unacquainted with the realities of modern air warfare. Balfour believed that he dealt humanely with the LMF cases that came to his attention, because, as he recalled, his "sympathies always started off with the poor fellow." Balfour added that "I knew only too well myself what it was like to be scared stiff in air warfare."[39]

At the Ottawa conference Power planned to make clear his dissatisfaction with the British handling of certain personnel issues involving Canadian flyers, such as delays in the awarding of aircrew commissions to Canadians serving with the RAF, and the Canadianization of selected RCAF squadrons. As historian W.A.B. Douglas has remarked, feelings were running high, for personnel issues were "important in terms of Canadian self-esteem and identity," and Power thought that the RAF had not honoured some of its undertakings.[40]

As the man most familiar with the RCAF's difficulties in Britain, Air Vice-Marshal Edwards was summoned from his RCAF post in London to participate in the conference. Described as "a firm and sometimes intemperate proponent of Canadianization, ... he announced that his main objective was to elaborate on the 'many difficulties' he had encountered as he endeavoured to carry out the policies of the Canadian government."[41] One of the items on the conference agenda that Edwards and other Canadian delegates saw as bearing on Canadian sovereignty and self-esteem was the treatment of Canadian aviators removed from flying duties. Before the conference began, Edwards explained his views on LMF procedures to a special meeting of the RCAF Air Council held on 22 May 1942. Replying to the MNDA's concerns about airmen "taken off flying duties for disciplinary reasons or medical unfitness" being classified as waverers, Edwards pointed out that to remedy the situation the RCAF would need to set a "definite policy" in such matters. He agreed with Power that "a man 'washed out' from flying duties by reasons other than cowardice should be placed in employment suitable to his [former aircrew] category and temperament, rather than discharged." The members of the Air Council concurred that "equitable treatment" should be given to men who washed out during training, "and that the term 'waverers' should be used only to apply to personnel who have shown unwillingness to fly." At this point there was no quarrel with the British handling of "straightforward" waverer cases or of those who were considered cowards. However, the Canadian air force officers did object to the RAF's treatment of most airmen charged with misconduct, of aircrew trainees who failed their course for reasons beyond the trainees' control; and of "medical waverers."[42]

During the second half of the Ottawa Air Training Conference, Power, Balfour, and their advisers entered into discussions meant to address the remaining issues related to LMF policy, the most important and contentious being the reasons for removal of flying badges. The Canadians insisted that for a man to lose his badge he must be "proved *wilfully* inefficient and/or unsatisfactory ... before being employed in full operational duties."[43] The criterion of wilfulness was the Canadian answer to the problem of how to distinguish between aviators who deliberately refused to fly and those who, while unsuited for operational duty, should be held blameless.

British Air Ministry regulations, however, required that the badge be removed, without exception, from anyone "withdrawn from flying duties before he has been employed on full flying duties."[44] This policy was an uncompromising one, but it was not illogical. It was based on certain principles that the Air Ministry did not wish to concede, and was rooted in a desire to preserve the morale of serving airmen. The Air Ministry did not want to devalue the badge by allowing it to be worn by AC2s cleaning latrines. The Air Ministry also wished to penalize aircrew candidates who failed training, as a means of discouraging others who might be thinking of joining the air force to elude danger in the army or navy. The Air Ministry also believed that allowing men who failed aircrew training to remuster within the RAF would leave the recruiting system open to abuse, for those anxious to avoid combat service would join the RAF and deliberately fail flying training to get a safe ground job.[45]

The more lenient Canadian approach toward the issue of badge removal can be explained in part by the fact that Canada's manpower situation at the time of the Ottawa conference was not as serious as Britain's. Unlike Britain, Canada still had an untapped pool of potential recruits to draw on. The Canadians could thus afford to take the position that it was unfair to deprive a man of insignia he had already won. Because, under the BCATP, wings were conferred with "due ceremonial"[46] at the completion of training in Canada, and their award was a "distinction not lightly bestowed," it was considered "a serious step" to remove them. In the RCAF wings were earned when one reached a certain stage in flying training, usually completion of SFTS. The British granted them at the same time, but "in anticipation of successful employment in an operational unit."[47] In the Canadian view, wings represented a training qualification; in the RAF they were awarded provisionally, and could be retained only by those who were able to fly operationally. The RAF attached no special significance to the badge before operational duties had been undertaken, and in contrast to Canadian practice, badges were often distributed, without formality, from a cardboard box in the classroom.[48]

In the end, the British and Canadian delegates to the Ottawa conference remained unconvinced by each other's arguments, and it was left to Edwards and Babington to sort out the aircrew-disposal issues on their own. After the Ottawa conference, Babington returned to London and told his staff that "the Canadians were most concerned about the removal of the flying badges from men who became unfit at OTUs through no fault of their own." Three weeks later, however, he recorded that he was "none too clear" on what changes the Dominion representatives wanted made to "the w procedure." Further discussions between the RAF and the Dominion air forces revealed that the RCAF and RAAF were now, in addition to their previous objections, opposed to the practice of keeping waverers at the rank of AC2 for three months.[49] To discuss these differences a meeting was convened at the Air Ministry in London toward the end of July, this time chaired by the AMP himself; Canada was represented by Edwards. The minutes of this meeting indicate that the Dominions accepted the RAF's view that medical waverers "were not purely medical cases," and could be dealt with under the LMF Memorandum. This was quite different from what Edwards had told the RCAF Air Council in May. He is also recorded in the minutes as stating that he had no objections to the practice of keeping ws at the rank of AC2 for three months. In return for these apparent concessions, Babington attempted to meet Dominion concerns by conceding that OTU trainees who washed out for reasons beyond their own control would "normally" be allowed to keep their wings. He also agreed to ask the Air Council to amend its confidential regulations so that airmen punished for "inefficiency" or for breaches of discipline would not automatically lose their wings, but would have their cases decided on the merits.[50] This accorded with the Canadian view, as expressed at the Ottawa conference, that wings should not be withdrawn unless the inefficiency or misconduct was deliberately intended to avoid operations.[51] Two months after this latest meeting, Edwards informed Canada that final agreement was near, and that he would interview all LMF cases before their fate was decided.[52]

However, not all the Dominion representatives were so sanguine. Shortly after the meeting, the Australian representative, Air Vice-Marshal F.H. McNamara, complained that he had not agreed to accept the RAF's medical waverer and AC2 policies, and that the minutes failed to record his objections. Air Ministry officials questioned McNamara's version of events, and explained that Babington, with the secretary of state for air's sanction, had decided that because the purpose of the LMF Memorandum was "the general maintenance of morale," it was "essential" to apply the Memorandum to medical waverers.[53] At the same time, however, because "the manpower situation [was] not so favourable," the Air Ministry did agree with the RAAF's suggestion that the requirement that AC2s be held back from remustering and training for three months be dropped.[54]

A compromise seems to have been effected when new Air Ministry regulations were issued in February 1943. These acknowledged the principle that aircrew badges were only to be withdrawn in cases where the individual was removed from flying duties for "reasons within his own control," or if he was classified LMF. Other cases were to be decided on their merits. But the RAF's medical waverer procedures were to stand.[55] After negotiations that had lasted over a year, matters seemed to be settled.

It is plain that there was considerable confusion among the participants over the Canadian position during the course of the negotiations that led to the compromise of February 1943. This is not surprising given the large number of officials, from the MNDA to junior staff officers, who spoke for Canada. Moreover, the Canadian position evolved over time, and in some instances RCAF staff officers in Britain were not consulted before changes were made by Ottawa. The files of the period reveal also that there was a striking lack of consistency at RCAF Overseas Headquarters in assigning staff to the negotiations. This may have been because Edwards was ill throughout his time in England,[56] and because he did not have confidence in his staff. Edwards complained that the "worst officers" had been sent to RCAF Overseas Headquarters, and that his assumption of command in November 1941 had not been popular. His arrival had created a "hostile atmosphere" among his staff, and he claimed that "everybody [was] diametrically opposed to policies from Canada."[57] Despite these problems, Edwards maintained at the beginning of

October 1942 that the problem of waverer cases would soon be "fairly well cleared up."[58]

Power, however, was concerned that there were no formal regulations "governing personnel, Overseas or in Canada, removed from Aircrew Duties for Misconduct, Inefficiency, or Course Failure." He was also troubled by stories of unfairness to alleged RCAF waverers repatriated to Canada, and consequently turned to his Air Council for advice.[59] Power and the Air Council concluded that specifically Canadian regulations were required, and action was taken in November and December 1942 to create them.

A lawyer by training, Power was personally involved in formulating the new regulations. But his was not the only legal mind at work. Beginning with the Ottawa Air Training Conference, Canadian drafts of the new regulations had been revised by the RCAF Legal Branch.[60] This marked the arrival of another technocratic élite, the legal profession, into the LMF debate, and they were to make their influence felt by introducing certain legal concepts into the new regulations. Senior Canadian officials in the Ministry of National Defence also influenced the evolution of the policies that affected airmen who were labelled waverers.

Most members of the RCAF Air Council shared a number of perceptions that affected the drafting of the new regulations. The first was that British treatment of some Canadian aviators branded LMF had been unduly harsh, as in the case where a "man [had] been returned to Canada, [and] his badge removed, after numerous operational tours." Secondly, there was a feeling that British opinions were being accepted uncritically by some RCAF officers. The MND, who had reviewed some of the files of those judged to be waverers, "made pointed reference to the rubber stamp nature" of the Canadian AFHQ board that reviewed these cases, and "suggested that this Board should submit its opinion independently of [British] views." Thirdly, the Air Council believed that the term waverer was being applied too broadly by the RAF, and that it should be restricted to flyers who wilfully shirked their duties. Finally, the Air Council asserted that no one should be deprived of his flying badge except "in cases of absolutely proven cowardice."[61]

After lengthy discussions, Power approved, in December 1942, new regulations governing deprivation of the flying badge and the treatment of aircrew removed from flying duties. These stated that "when it was clearly established" that an airman was "guilty of a 'Wilful' attempt to escape from the hazards of operational or other Aircrew duties," for reasons of "'lack of confidence or moral fibre,' misconduct, inefficiency or course failure," he was to be "deprived of the flying badge" and "discharged or retired" from the air force. On the other hand, if, on the basis of the airman's service record and medical history, including his history of any combat operations or accidents, it was "not clearly established" that there had been a wilful attempt to escape aircrew duties, the airman was to be permitted to "retain the flying badge" and was to be "retained in the Service," if that was "in the best interests of the Service"; otherwise, he was to be "discharged or retired."[62]

By applying the test of wilful avoidance to all aircrew removed from flying duties, not just to those who failed at OTUs, the Canadian regulations departed significantly from previous British and Dominion policy. They advanced the notion of volition, as a key concept, to centre stage in the LMF debate. This would lead to a major divergence between how the RCAF and the RAF dealt with aircrew who were removed from flying duties.

The regulations were designed to assert Canada's responsibility for all of its nationals who were suspended from flying, and to place final decision-making authority in the hands of the MNDA. Power denied the request of his service advisers that the right to deprive aircrew of flying badges be retained by the Air Officer Commanding-in-Chief (AOC-in-C) RCAF Overseas.[63] Instead, he instituted a detailed assessment process that provided for the participation of Canadian officers at each level, and culminated in ministerial review of each case. Once a man had been removed from aircrew duties, and the RAF procedures begun, an RCAF Overseas Headquarters board was to "investigate and review" the case "prior to the decision of the Air Ministry." This board was to submit its findings "through the usual channels to [the] Air Ministry." All of the documentation was then to be presented to the AOC-in-C RCAF Overseas, who was to forward his recommendations to air force headquarters in Ottawa. He could also recommend that the airman be repatriated to Canada "pending the final decision of the Minister." In the meantime, all of the cases of those who were repatriated were to be reassessed by an AFHQ board, which was to submit "its own findings and recommendations to the Minister." Those who remained in Britain were to be dealt with under different procedures (described later in this chapter). But in every case "[n]o action by way of reduction in rank, remustering or deprivation of flying badge" was to be taken until the MNDA had rendered his final decision.[64]

This seemed to many senior officers to be a fair, if lengthy, procedure, but not all agreed. Sitting far away in London, Edwards was taken by surprise by the new regulations. In January 1943, just as he was fine-tuning the arrangement he had made with the British Air

Ministry, he was amazed "to see the [RCAF] Air Council Minutes wherein it was decided that waverer cases should be handled in Canada." While protesting that the process he had instituted in Britain was "established" and "sufficient," and that he should have been consulted before any changes were made, Edwards announced that he would "fall in line," but predicted that the new approach would cause problems.[65]

At the beginning of March, the news was broken, to what must have been an equally surprised British Air Ministry, "that a recent meeting of the Air Council (RCAF) has ruled that deprivation of the flying badge will only take place in Canada under the authority of the Minister of National Defence for Air ... RCAF aircrew will not be deprived of the right to wear the flying badge while in the United Kingdom."[66] Confusion arose several weeks later when Air Commodore F.G. Wait, the new Director of Personnel at RCAF Overseas Headquarters in London, informed Air Ministry officials that no changes would be made to the way reductions in rank were handled, except in cases governed by the LMF Memorandum.[67] The RCAF Air Council's instructions, however, were clearly to the contrary.

The last act in what was turning into a bureaucratic farce occurred at a meeting in London in March 1943 called for the purpose of amending the LMF Memorandum. The meeting was chaired by the new RAF AMP, Air Marshal Sir B.E. Sutton. Canada was represented by two RCAF squadron leaders,[68] while the other Dominions were represented by officers of air (general officer) rank. Aside from some minor clarifications to the categories of waverers, it was decided to continue with the RAF procedure for handling medical waverers, which meant that removal of the flying badge and reduction in rank were both to take place in Britain. Once again RCAF officers in Britain appear to have acquiesced to a policy different from the one approved by RCAF authorities in Canada.[69] Perhaps Canada should have sent some officers of a higher rank than squadron leader to the meeting, and ones more in tune with Ottawa's wishes.

This was one more example of the confusion that reigned throughout the long negotiations between Britain and the Dominions. The Dominion representatives believed that British Air Ministry bureaucrats had deliberately obstructed their efforts to effect policy changes. The Canadians and Australians found evidence of this in allegedly intentional misrepresentations of Dominion policies by the British.[70] To the British, however, it seemed that the Dominions changed their policies on a daily basis, and in the RCAF's case that policies varied depending on whom one was dealing with. In the sixteen-month period from December 1941 to April 1943, at least eight RCAF Overseas Headquarters officers attended meetings with or wrote letters to the Air Ministry in connection with the negotiations,[71] and they frequently contradicted each other and AFHQ in Ottawa.

Another reason for the differences between the RAF and the RCAF stemmed from the ways in which each service functioned on a day-to-day basis. Before the negotiations began, both were somewhat independent of ministerial control, and their respective Air Councils prepared almost all of the regulations and directives that governed each air force. The responsible minister, in Britain the secretary of state for air and in Canada the MND, "acted as a kind of chairman of the board," and until 1940 the Canadian minister was seen as a "mouthpiece" for the views of the service members of the Air Council. Once Power became MNDA, he reduced the Air Council's authority and asserted his control over it. Power believed that the Air Council "had assumed to itself powers of administration and direction far beyond" what he could countenance, and he made it clear to its members that they "must confine themselves to the duty of making recommendations to the minister." Although Power noted that it "took some time before they fully accepted the idea," the Air Council under Power was, in theory, much more responsive to political direction than its RAF counterpart.[72]

The records of the waverer discussions reflect the new state of affairs under Power. We can see that, even though Power consulted with the Air Council, he was the one who controlled the agenda. He also participated actively in the Air Council's discussions and insisted on making the final decisions.[73] This was in contrast to the passive British secretary of state for air, Sir Archibald Sinclair, who was rarely involved in the discussions, and when he was, relied heavily on the counsel of his senior service advisers.[74]

But differences in administrative procedures were not the only reason that the RCAF and the RAF differed in their approach toward sensitive personnel issues. A number of other issues regarding how alleged waverers were being handled soon surfaced. Not long after the LMF policies were revised in mid-1943, they became the target of criticism. The president of the RCAF board that was established to deal with LMF cases noted that by July (the month after the LMF Memorandum itself had been amended) the "executive action being taken by the RAF [had] apparently deteriorated in quality and effectiveness," and that the amended document needed review.[75] His assessment was shared by the RAF's consultants in neuropsychiatry. They explained that "there [was] considerable uncertainty about the meaning of the term lack of moral fibre," and stated that COS were not distinguishing between LMF cases where the airmen had tried their best but broke down because of a predisposition, and LMF cases where the airmen had not demonstrated any effort at all to rise to the demands of service life. They offered many suggestions on modifying the Memorandum,[76] but none were accepted until its final revision in March 1945, when the RAF accepted the advice of its neuropsychiatrists and acknowledged the point the Dominions had been making all along, namely that some people who could not be fully employed operationally might still deserve to keep their wings.[77]

Most of the difficulties between the RAF and the Dominions over the treatment of aircrew who were removed from flying duties revolved around the handling of NCOs, who made up most of the cases.[78] As societies that were more egalitarian than Britain, the Dominions found it difficult to countenance the fact that non-commissioned personnel were sometimes treated more harshly than officers, especially in W cases.[79] The differences in treatment have been commented on by a number of authors. Terraine has calculated that of all the LMF allegations brought to its attention, the Air Ministry officially confirmed 70 per cent of the cases involving NCO aircrew, but only 52 per cent of the cases involving officers. In Terraine's opinion, the dice were loaded against NCOs.[80] Hastings has noted the RAF's reluctance to take action against officers holding permanent commissions, because a court martial was usually required to deprive them of their rank.[81] As we have seen, a court martial, in Bomber Command at least, was very difficult to arrange, and court martial or not, a recommendation for terminating an officer's commission had to go all the way to the secretary of state for air for approval.[82] To avoid this long process, the most expeditious means for dealing with officers suspected of LMF was often a quick posting to a new assignment. But for NCOs believed to be waverers, a number of actions, not always conforming to official policy, could be taken at the unit level. Some NCOs were sent to "Aircrew Refresher Centres," which were really open-arrest detention barracks where a rigorous regime of physical training was enforced, and which by 1943 were "handling thousands per year."[83] Or an NCO might be publicly humiliated by having his rank insignia and flying badge removed on a station parade.[84] Alternatively, he could be deprived of rank insignia and flying badge at a holding depot and then be posted to a remote place.

The lack of procedural safeguards for RAF NCOs, and the willingness of the RAF to treat officers and men differently during World War II can be explained in part by the organizational culture of the service. During the interwar period, RAF officers with permanent commissions – the leaders and policymakers of World War II – were exhorted, in the words of one wing commander, "'to inculcate a public school feeling'" in their units. The RAF deliberately cultivated this model of behaviour during the 1920s and 1930s to make up for its perceived social inferiority to the other two services, particularly the army.[85] A senior RAF officer wrote shortly after World War II that

The glory of the public school system in the past and its prime justification today is that it is the best-known method of producing leaders of men. It is unfashionable today to talk about an Officer Class. No one will be so bold or foolish to deny that, in horses, breeding and training are indispensable if one wants to produce winners; and I have never been able to make out why anyone should think that does not apply equally to men ... But if we believe in the public school system, if we continue to claim its privileges ... let us admit that it does, and I believe always will, produce a very high proportion of the best leaders of men in Britain.[86]

Even if this view was, as Lesley Fernandez-Armesto has claimed, "romanticised and evidently false,"[87] RAF officers, especially those living in Officers' Messes, were encouraged to behave like gentry.[88]

British NCO aircrew, on the other hand, were treated as members of a lower class. The expanded RAF of the later 1930s and World War II years was not prepared to commission all pilots. Those pilots who were deemed not to be officer material were made sergeants. During the interwar years most aircrew positions other than pilot were filled by junior ranks, usually maintenance personnel, as a secondary duty, and early in the war it was still not uncommon for leading aircraft-men to act as air gunners.[89] Part-time aircrew were required to do their regular jobs, as well as station chores such as all-night guard duty, in addition to their flying.[90] On the ground they were subject to parade-square discipline. One unit, 82 (Bomber) Squadron, was described as "highly disciplined" because "[Warrant Officer]] Paisley the Squadron disciplinarian saw to it that ... when '82' wasn't flying they were marching." In fact they drilled so much that they were referred to as "'Paisley's 82nd of Foot.'" This practice continued even while No. 2 Group, of which 82 Squadron was a unit, was suffering huge losses in the air.[91]

During the early part of World War II, as aircrew tasks became more specialized and highly regarded, aircrew were usually excused from station duties, and all were given the rank of sergeant or higher. Eventually all aircrew positions became full-time. Neverthe-less, intolerance of NCO aircrew persisted. In 1940 Wing Commander R.F.T. Grace, a neuropsychiatrist, noted that 50 per cent of aircrew officers but less than 25 per cent of sergeant aircrew who experienced psychological disorders returned to flying duties. He also discovered that "the sergeants appeared to break down more completely than the officers." Grace "wondered if this was due to the higher morale of the commissioned officers." What he did not say was that better morale might have been the result of superior living conditions, minimal harassment from station administrators, and the fact that officers were generally held in higher esteem inside and outside the service. The problem of inadequate amenities for NCOs was recognized by the DGMS Committee on Flying Stress early in the war. One MO suggested all-ranks Flying Crew Clubs, reasoning that young flyers, whether officers or NCOs, did not mix well with the older generation of RAF personnel.[92] His was a valid point, for serving aircrew did grouse about the fact that they were always risking their lives while older, more senior men stayed behind and avidly pursued career advancement.[93] The older generation, on the other hand, saw a group of young men who had attained officer or senior NCO status and privileges simply by virtue of serving as aircrew. In their minds these youngsters were undisciplined and ignorant of RAF traditions.[94]

If British NCO aircrew were generally treated poorly in comparison to their officers, Canadian NCO aircrew were victims of double jeopardy. As "colonials" they were often seen by the British to be lacking in manners and to know nothing of the deferential attitude expected by their superiors. In 1941 tensions between the Canadians and the British led to the "Cranwell Mutiny," a strike by the first group of Canadian wireless operator (air gunner) sergeants to arrive in Britain for final technical and OTU training. Instead of receiving operational training, the Canadians were sent to the cadet college at Cranwell and subjected to route marches and parades as if they were raw recruits. They reacted by ignoring their RAF superiors' orders and finally by throwing all their barrack's furniture onto the parade square and setting it on fire. Their revolt ended when the Canadian High Commissioner in London intervened to ensure that they were posted out and given the expected instruction.[95] This incident illustrated what Power feared could happen to "well educated citizens of the Dominion" when they were not under RCAF supervision.[96] In the eyes of some RAF personnel, however, Canadian NCO aircrew were just "undisciplined savages."[97]

Power's solution to this dilemma was to gradually increase the percentage of Canadian aircrew who were made officers, and ultimately to grant officers' commissions to all Canadian aircrew. It did not seem right to many Canadians that NCOs doing the same jobs as officers should live in inferior conditions and receive just half the pay.[98] According to Power, every crew member shared the same risks, and therefore all were entitled to the same rank and pay. He used the analogy of an English cricket match to explain his opposition to the unequal treatment given to officers and NCOs. Power remarked how ridiculous it was that "at the adjournment for tea after playing together most of the day some of the team repaired to a tent marked 'Gentlemen,' and others to a tent marked 'Players,'" even though minutes before all had been working together as members of the same team.[99] The Canadian view has been characterized by Terraine as having "logic and humanity on its side," and as being more appropriate to a large, technical service such as the air force, where talent, not class, was all that supposedly mattered.

The Canadian desire to treat non-commissioned personnel with consideration was apparently also shared by Hugh Trenchard, "the father of the RAF." His biographer tells us that in 1916, when Trenchard was GOC of the RFC in France, he criticized one officer's excessive punishment of some mechanics with the remark that "'This is a technical corps. Our job is to shorten the war. You're not in the army now, you know.'" Trenchard's ideas were generally accepted in the RAF between the wars, but at the beginning of World War II there were still air force officers who valued discipline over technical proficiency.[100]

The RAF and the RCAF, however, did not see eye to eye on the issue of increasing the ratio of aircrew officers,[101] and in any event the scheme could not be implemented quickly. As William Carter has pointed out, there were a number of obstacles to carrying out Power's plan, including the RAF's reluctance to have different officer-selection criteria among Commonwealth aircrew who would be flying together, and such practical considerations as insufficient messing facilities for the large number of new officers who would be created under the Canadian scheme.[102]

It is true, however, that the British attitude toward Canadian airmen was not entirely without foundation. One problem was that the newcomers were unfamiliar with service traditions and untrained in leadership skills. In January 1942 Edwards complained that the "discipline of our troops in England [is] tragic," and he attributed this to the fact that RCAF aircrew had "not been taught in Canada what it means to be an officer or NCO." He emphasized that *flying, technical and academic training in itself [was] not enough*"; leadership training was also required, but no thought had been given to it by the Canadian authorities despite his "vigorous protests."[103] Edwards also remarked that, in contrast to the behaviour that had led to the Cranwell Mutiny, some senior RAF personnel had treated Canadian airmen guilty of misdemeanours "more like guests than culprits." To rectify the discipline problem, senior RCAF officers agreed in March 1942 that "firm disciplinary action ... on precisely the same footing as British personnel" was now required.[104]

But disciplinary action alone could not compensate for a lack of leadership training. According to the RCAF Air Council "the need for ... [leadership] training was very apparent" as late as 1943. Nevertheless, "time could not be spared for this purpose," and the "haphazard" system "in which ... newly commissioned officers [were] left to acquire the necessary training by observation only" was left in place. The problem was not unique to the Canadian air force, moreover. The RAF, which like the RCAF had also experienced an influx of newly commissioned officers in 1943, arranged for leadership training in the United Kingdom, but could not help the RCAF, as "facilities there [were] adequate only for RAF personnel."[105] Yet it is not certain that leadership training, if given in Canada, would have had a lasting effect on RCAF aircrew, as it had been noted that the "strict discipline" which was enforced at the Canadian BCATP schools was relaxed once they arrived in Britain.[106] In addition, long waits at RAF holding depots, where the lack of activity made everyone "bone lazy," contributed to a decline in standards of dress and deportment.[107]

Aside from the absence of leadership training, there were other reasons why the RAF, especially Bomber Command, was concerned about the large influx of Dominion aircrew. As the percentage of "colonials" in Bomber Command increased, it was feared that the force was not taking in enough young men of the British middle and upper classes, the supposed natural leaders and "backbone" of British society.[108] Although Harris was not a supporter of "the old school tie" as the prime criterion for granting an officer's commission, he was alarmed by the "alienization" of the RAF, and for good reason. By January 1943 almost one-third of his squadrons were designated Dominion or Allied, and Canadians eventually composed about 25 per cent of Bomber Command's aircrews.[109] Harris worried that if "'most of the operational squadrons were manned by coloured troops,'" as he disparagingly referred to Dominion aircrew, he would lose operational control of his force, because the Dominions insisted on being consulted on personnel issues such as aircrew disposal and tour length. To hold on to what he saw as his operational prerogatives, Harris sometimes acted independently of Air Ministry policies.[110] This, and other RAF actions that undermined Canadian authority, eventually forced the RCAF to take measures to ensure that its wishes were respected.

One way the RCAF tried to gain some measure of administrative jurisdiction over its personnel was to establish a unit in Britain where aircrew who were medically unfit or suspended from duty could be gathered together under RCAF control while their cases were being considered. Immediately after his arrival in London, Edwards had decided that it was essential to establish a "disposal centre for unemployable RCAF personnel," including medical cases and alleged waverers, so that as many as possible could be "reclaimed for service" overseas.[111] However, it was to take almost a year to organize a records system to keep track of Canadian aircrew overseas,[112] and to find and prepare a suitable location for the disposal centre. Until this was done, "unemployable" RCAF aircrew were scattered throughout England while they waited for repatriation or to have their cases heard. Such waits could last up to eight months.[113]

With RAF expansion and the increased intensity of operations in 1942 the number of aircrew requiring processing grew, and the RAF was forced to open two new holding and reselection units at Brighton and Blackpool in the fall of that year for aviators "suspended from air training or operations." All WS were, in theory, to remain segregated at Uxbridge, but some were, to the detriment of morale, mixed in with men awaiting retraining assignments or medical board hearings at the new locations.[114]

Even though the RAF depots had the capacity to handle all Canadian aircrew, the RCAF pressed ahead with the establishment of its own disposal centre. The RCAF's "R" Depot opened in the fall of 1942 at Warrington in Cheshire, but inadequate facilities and the small number of aircrew requiring processing meant that initially only medical boards were conducted there.[115] In the meantime, RCAF aircrew who failed training were sent to the RAF reselections board at Brighton, while waverers proceeded to Uxbridge, as before.[116] Most aircrew were then returned to Canada for final disposal of their cases. To look after Canadian interests, RCAF officers sat on the boards at Brighton during hearings involving RCAF personnel, but their recommendations were not sent to RCAF Overseas Headquarters unless the RCAF representative believed that a case had not been "fairly dealt with."[117]

This arrangement was not satisfactory to the CO of the Warrington depot, Wing Commander (later Group Captain) Denton Massey.[118] In early 1943 Massey wrote to Edwards charging that the proper operation of the depot's medical boards was being frustrated by deficiencies in staff and facilities, by a poorly defined medical policy, and by their reliance on British reselection boards to ratify RCAF findings. In addition, the practice of repatriating almost all cases to Canada for final disposal had significant drawbacks. The long distances involved, combined with poor record-handling, left the board in Canada (located at RCAF Station Rockcliffe) with no first-hand access to information on the reasons for the suspension from flying duties of Canadian airmen in Britain. At the stage of final disposal this resulted in delays and erroneous decisions, and caused an avoidable waste of manpower.[119] Massey wanted more than just medical boards at Warrington, and he urged that autonomous re-selection boards (which would make use of the medical boards already at Warrington) be established there, with the power to dispose of cases without having to seek British approval. He believed that if reselection could be managed by the RCAF in Britain, some aircrew could be quickly and usefully employed somewhere overseas, and that a "considerable salvaging" of manpower would result.[120]

In answer to Massey's criticisms, an RCAF reselection board was instituted at Warrington in May 1943, although at first its findings were sent to the RAF reselection centre at Eastchurch for approval. One month later, when RCAF Overseas Headquarters informed the British Air Ministry that it intended to establish an independent reselection board at Warrington, the RAF objected on the grounds that it would duplicate existing RAF facilities, and could lead to inconsistent treatment of airmen suspended from flying duties.[121] Edwards ignored British demands for written confirmation of his plans, and one day before the RCAF reselection board began its work, his staff informed the RAF that the services of its board at Eastchurch would no longer be required, and that the "R" Depot would deal directly with the Air Ministry. From 1 October 1943 on, the RCAF operated an autonomous reselection board at Warrington, and this enabled Canada to increase its administrative control over RCAF personnel.[122]

Senior RCAF officials had agreed in early 1943 to discontinue sending their W cases to Uxbridge, and as soon as the reselection board was functioning at Warrington the Ws were sent there. In March 1944, with prodding from AFHQ in Ottawa, a Special Cases Committee, separate from the reselection board, was appointed to process alleged waverers. Even though the reselection board and the Special Cases Committee were composed of the same officers, it was felt that a separate body was needed to handle LMF cases.[123] Beginning in April, the boards at Rockcliffe were discontinued, and recommendations for the disposal of all cases were forwarded directly to the MNDA from the Canadian boards sitting in Britain. Airmen now awaited the minister's decision at Warrington, and this allowed for speedy reallocation of those who were still considered employable outside Canada. In September 1944, two months before he resigned from Cabinet over the conscription crisis, Power disengaged himself from the process, and directed that the AOC-in-C RCAF Overseas would now be the final authority in all cases dealt with at the "R" Depot.[124]

As the Warrington operation grew in size and scope, significant differences emerged between it and its RAF analogues. The first was the relative autonomy of the Canadian panels from headquarters control. Edwards expressed his "full confidence" in the Warrington boards, and the CO of the "R" Depot was allowed to "make a decision in each case and ... take action upon ... [his] own authority" without reference to RCAF Overseas Headquarters.[125] Recommendations were then forwarded to the appropriate authority for final approval. In the British air force, by contrast, all authority for the disposal of cases involving aircrew suspended from duty rested with the Air Ministry. COs of RAF holding depots were relegated to routine administration, and it did not "seem to be their business to concern themselves with the action taken (or not taken)" in the cases processed at their units.[126] Another major difference was the Canadian emphasis on reassigning personnel as quickly as possible;[127] whereas long waits were the norm in RAF cases, Warrington aimed to process people in weeks rather than months.[128] The final difference was the RCAF's adoption of its own alternatives for the disposal of cases to replace those in the LMF Memorandum. The Special Cases Committee assigned aircrew to one of seven categories, compared to the RAF's three, to allow for more precise and, it was hoped, fairer dispositions.[129] (Appendix B provides a statistical breakdown of how cases were handled by the Special Cases Committee during a one-year period in 1944–45.)

The new Canadian boards and committees helped improve disposal procedures once cases reached them, but a weak link remained at the unit level. Contrary to British Air Ministry directives, as late as February 1943 some airmen were being sent to Warrington already reduced in rank and deprived of their flying badge.[130] At the end of 1943 Massey protested that "LMF'ing in many instances is being proceeded with loosely and without due regard to the seriousness of the action."[131] From at least 1942 on, administrative chaos and other problems had resulted in an "extreme time lag" in many cases. An unwieldy chain of command (four levels of intermediate authority interposed between squadron COs and RCAF Overseas Headquarters) was responsible for delays of "months" in communications between Edwards and the squadrons. This exacerbated the difficulties of implementing complex and changing policies. Instructions were still being issued in April 1944 by No. 6 (RCAF) Group Headquarters "to clarify the confusion which would seem to have arisen with regard to the disposal of aircrew personnel in this Group." The situation had not improved much by November, as there was still "a considerable delay in taking the executive action necessary."[132]

One of the difficulties in carrying out aircrew-disposal policies was the administrative burden placed on individual squadrons. Understaffed and sometimes equipped with only one typewriter, squadron orderly rooms were unable to cope with rapidly increasing administrative demands, and yet they were expected to comply "strictly with the provisions of the [LMF] memorandum."[133] In addition to the voluminous routine statistics, reports, and returns that had to be completed, the paperwork associated with cases suspended from flying was onerous. As we have seen, LMF policies changed frequently, and as each new one was promulgated, it had to be digested, interpreted, and implemented. Instructions issued to No. 6 Group squadrons, to offer an example of the paperwork burden, required for each case that all documents be prepared in triplicate, with all certificates properly completed, signed, and initialled, all medical reports attached, and a "list of sorties (if any) to be shown separately, with date, place, duration, and remarks as to whether successful, any difficulties encountered, etc."[134] While this was deemed necessary for the proper consideration of each case, it shifted an enormous administrative workload onto those least equipped to handle it. This may also help to explain why COs preferred posting, or other less complex administrative actions, to the LMF procedure.[135] As fairness became more of a concern, and the information required by the boards increased significantly, it became difficult to achieve a balance between bureaucratic requirements and the ability of the squadrons to cope with the avalanche of paper work. Compromises had to be made, and sometimes they undermined the goals of Canadianization. In April 1944, for example, the AOC-in-C RCAF Overseas found it necessary, because of administrative delays to delegate "[r]eduction of rank, remustering for inefficiency, reclassification, etc." and court martial authority to the RAF Commands.[136] His decision, of course, ran counter to RCAF policy, and may have adversely affected the morale of Canadians serving with the RAF.

Canadian morale was also at risk for other reasons. Sir Arthur Harris was convinced that "'morale was never a problem'" in Bomber Command,[137] but, as we have seen, the resolve of the men engaged in the bomber offensive was of constant concern to senior air force officers. The Air Ministry was intent on keeping casualty figures a "closely guarded secret throughout the war," as their publication would have had a profound effect on morale in Bomber Command.[138] The question of how to maintain fighting spirit gave rise to a number of studies, among them a survey of RCAF personnel conducted in late 1942. The Canadian survey concluded, as had British researchers, that "the problem of the Waverer is intimately related to the morale of the fighting unit." If unit spirit was high it prevented "cracked nerves" among aircrew and deterred shirking. The authors of the survey believed that in units with good morale most airmen preferred death to dishonour, and would stick with their comrades rather than "go LMF," as aircrew said.[139]

Morale is a difficult attribute to quantify in any fighting unit. In air combat it has been linked to many indices, including casualty totals, early-return and non-starter rates, LMF rates, and NP ratios. Historians of No. 6 Group have suggested that re-equipment programs, inexperienced leaders and crews, inadequate aircraft maintenance, and high casualties all contributed to low morale, particularly during its first year of operations.[140] Edwards believed, based on his own observations and on waverer statistics collected from five Groups in Bomber Command during the months of June to August 1943, that No. 6 Group's morale was comparable to that of the RAF's bomber groups.[141]

Statistics compiled by the Special Cases Committee at Warrington from September 1944 to September 1945, which are set out in appendix B, make it possible to calculate an approximate NP ratio for all RCAF aircrew, including those who served in RAF units. As in chapter 5, the intent here is simply to establish an order of magnitude. During the period in question, 94 aircrew classified LMF or "Medical 'w'" arrived at Warrington. If the RCAF had the same ratio of LMF and Medical w cases to "neurosis" cases as existed in the RAF, there would have been 895 neurosis cases at Warrington in addition to the 94 LMF and Medical w cases, for a total of 989 cases. Of those, approximately 300 would have come from Bomber Command. The ratio of the total number of LMF, Medical w, and neurosis cases (300) to an adjusted annual RCAF aircrew casualty figure for Bomber Command during the period (4,334) yields an NP ratio of almost 7 per cent. Similar calculations for RAF aircrew in Bomber Command produce an NP ratio of about 6 per cent.[142] (Detailed calculations are in appendix C.) These figures suggest a comparable state of morale among RCAF and RAF aircrew in Bomber Command during the last phase of the war.

In the disposal of aircrew suspended from flying duties, as in selection and training, the RCAF, without much experience of its own before 1939, was compelled to adopt British methods early in the war. The purpose of the LMF procedure, as some saw it, was to deal with cases of cowardice without resorting to the cumbersome military justice system;[143] however, as the aircrew-disposal problem began to overwhelm the antiquated RAF administrative system, the LMF procedure was expanded to handle all types of aircrew who were unemployable in their assigned trades. Difficulties with the British LMF process became evident in Canada when the label of waverer was applied to some RCAF airmen who clearly did not deserve it.

This injustice caused political and social reverberations in both Britain and Canada, but in Canada the protests were more vocal and widespread. Public opinion in Canada demanded that the nation's cream of the crop be treated with respect and consideration, not like "savages" or "coloured troops." Power declared that his government "had a moral responsibility" for the welfare of Canadian youth serving with the RAF, especially those in the other ranks, who were particularly vulnerable to the caprices of the British disciplinary system.[144]

Once appointed MNDA, Power set in motion several modifications to the procedures for dealing with aircrew suspended from flying duties, but in effecting these changes the RCAF laboured under a number of handicaps. This produced a major rift between the RAF and the RCAF in their attitudes toward LMF.

British LMF policies were an amalgam of bureaucratic expediency, psychological theories, and operational imperatives. This was to be expected, as their authors included Air Ministry bureaucrats, medical specialists, and staff officers, all subject to the influence of the operational Commands, especially Bomber Command. In contrast, the RCAF under the guidance of Power, a lawyer by training, devised an LMF procedure that was primarily legal in nature.[145] Drafted with the assistance of air force lawyers to achieve Canadian political objectives, the RCAF's LMF regulations emphasized due process and the protection of individual rights. In adopting the criterion of clear and wilful evasion of operational responsibility as the basis for judging the behaviour of aircrew, the RCAF regulations moved the LMF procedure away from the bureaucratic, operational, and medical realms toward the political and legal arenas. At the same time, the Canadians provided for a more effective use of manpower than the British by ensuring, whenever possible, that aviators unable to cope with the rigours of operational flying were assigned other tasks within the air force.[146]

To those in Britain, enduring great hardships and under air attack for much of the war, Canada's insistence on individual rights and national sovereignty may have been hard to understand, particularly given the ineptitude of RCAF Overseas Headquarters staff officers in representing official policy. Opposition to RCAF initiatives at many levels of the RAF convinced Canadian authorities that the only sure way to accomplish their aims was to control the aircrew-disposal process as closely as operational necessity would allow. This led to the establishment of the RCAF centre at Warrington, where authority was vested in the CO to salvage as many airmen as possible. The Canadian system for disposing of cases of aircrew suspended from flying duties, which included an exhaustive review procedure whenever dismissal or disgrace was recommended, was noticeably different from that in use in the RAF.

In their attempts to achieve uniform treatment of LMF cases and to meet ever-changing wartime conditions, both the RCAF and the RAF introduced numerous alterations to the rules, regulations, and procedures governing alleged waverers. This kept squadrons that were already overloaded with paperwork constantly scrambling to keep up. Differences in administrative efficiency from squadron to squadron may have accounted for some of the variations in the way in which aircrew under suspension were treated. Squadrons that were able to handle the burden of their paperwork may have followed the regulations closely, while the others may have resorted to procedural expediencies. In the final analysis, the RCAF's complex scheme for managing LMF cases created a slow and cumbersome process that left many airmen to languish for months at holding depots, or to be dealt with summarily by operational units attempting to circumvent bureaucratic inertia. Manpower that could have been put to good use was thus needlessly dissipated.

In the British armed services, what were defined as operational priorities took precedence over many personnel matters until victory against the Axis was assured. The RAF Air Council, under pressure from Bomber Command, made judgments in areas, such as aircrew disposal, that were influenced more by operational necessity than by expert advice. However, as with aircrew selection, these judgments sometimes were responsible for manpower losses that could have been avoided. The final version of the LMF procedure, which made more efficient use of manpower than earlier versions, was the result of a desire by the government near the end of the war to avoid embarrassment. Despite opposition from some senior RAF officers, the Air Ministry, on being directed to do so by the government, finally accepted a view of LMF not too different from the one the Dominions had been advocating for much of the war.

But the RAF was only required to accept the Dominion view in late 1944. Before that time, protracted negotiations over the modification of disposal procedures achieved little from Canada's perspective, and the fundamental differences of outlook between the British and the Canadians could not be reconciled. By devising its own policies and administrative arrangements to achieve its political objectives in the LMF debate, Canada eventually asserted its independence from British control, and won a victory for national pride and political sovereignty.

7 None for All and One for One: The Canadian Manpower Crisis

The allocation of manpower in both world wars was of central importance to Canada's effort in those struggles and to its evolution as a nation. Conscription, with its inseparable military and political dimensions, was introduced in 1917 as a means of addressing a shortage of military manpower. From that time on, its potential to split the nation turned conscription into a "complex and contentious" issue that "haunted the minds of Canadian politicians."[1] The present chapter has as its main theme the RCAF's part in the Canadian conscription crises of World War II. The discussion here will assess the air force's role in these crises and its efficiency in managing its own manpower. To place RCAF actions in context, the chapter will summarize the main political events leading up to the conscription crises of World War II and compare the RCAF experience to the army's manpower dilemma.

On examining the Canadian conscription crises of this century it becomes clear that political considerations were exceptionally important. The military situation intruded from time to time on the political domain, but it was the government's assessment of the consequences of its actions – or inaction – in domestic affairs that determined how it would proceed. When discussing World War II manpower issues it is also important to remember that in Canada, as in other countries (including the totalitarian states), the allocation of human resources was a "delicate balance ... between compulsion, direction, persuasion, and voluntary responses," and that no nation actually waged "total war."[2] Under these circumstances the state could not employ too much coercion without risking losing more resources administering regulations than were gained by their implementation. Another important reason for minimizing compulsion was the need for society's backing to pay for war loans (such as victory bonds) and heavy taxes. As Canada discovered, "for any system of enforcement to work effectively, it must have the strong support of public opinion."[3] Canada also discovered that the most efficient means of gaining popular support included propaganda and financial and other incentives. Despite its widespread success in using those means to maintain Canadian support for the military effort in World War II, the Mackenzie King government faced manpower shortages that precipitated two conscription crises reminiscent of the crisis of World War I.

Heavy Canadian casualties on the western front during 1916 prompted the CGS to advise the prime minister, Sir Robert Borden, that conscription would be necessary to provide reinforcements for Canadian troops in France. After a great deal of political wrangling, the Military Service Act introduced conscription in October 1917, and was immediately attacked by many Quebec politicians. Laurier opposed it because it struck at "'the heart and soul'" of Confederation. Others in Quebec pointed out that they were willing to accept conscription for the defence of Canada itself, but would not recognize any government that imposed obligatory military service for imperial wars.[4] A mere 24,000 or so conscripts reached France before the war ended, which has led historians J.L. Granatstein and J.M. Hitsman to declare the Act "only a partial military success."[5] On the political front it created, in the words of Justice Lyman Duff, a legacy of "'hatred and bitterness'" that was a "'menace to national unity.'"[6] For this reason, Granatstein and Hitsman have called the introduction of conscription a "tragedy indeed, and a pointless one."[7]

Ronald Haycock has demonstrated that some of the blame, as well as some of the praise, for the army's manpower situation in World War I must be shared by Sam Hughes, minister of militia and defence from 1911 to 1916. His personal direction of the Canadian mobilization effort produced impressive initial results, but collapsed under the demands of a modern professional army.[8] Moreover, the government of the day was unable to decide how to apportion resources between the military and the "equally vital forces of industry and agriculture."[9] Problems of resource allocation were to return to plague Mackenzie King's government in World War II.

While the World War I experience provided many lessons for politicians and the army on how to put large forces in the field, there were few lessons of direct relevance to the air force. Because flying was more attractive than infantry service, RFC Canada had little trouble finding aircrew. In addition, "the heavy casualties among military aviators were not nearly so well known as those among subalterns in the CEF."[10] The recruiting of aviators seems to have had little effect on the army's needs in World War I, as the numbers who joined the flying services were tiny compared to what the land forces required. Of the over 600,000 Canadians in uniform in World War I, only 20,000 served in the flying services.[11] In fact, the militia did much to help the air force grow. The Militia Department informed the commander of RFC Canada that his organization was "free to enlist anybody" it could get (with some minor restrictions), and that it could "take any steps [it] lik[ed] to secure recruits." The Militia Department was "ready to co-operate with [him] to the fullest extent; and, practically, [he could] draw on the whole manpower of this country."[12] This was quite a different relationship between the air and ground forces from the one that was to evolve in World War II. The small size of the air forces of World War I gave rise to the idea (which still persists) that the aerial dimension of the war was insignificant, and required correspondingly few resources. By 1918, however, "one quarter of Britain's military budget ... was [being] spent on the air force," and one-half of the "contracts outstanding at the end of the war were air contracts."[13] Perhaps the rapid increase in the financial resources necessary to maintain the air arms at the end of the war, and the high casualty rates among airmen, should have alerted even those who were sceptical about air power that the cost of an air commitment, in lives and treasure, would be a great deal higher in future wars. Despite the omens, however, many politicians still believed that fighting the next war in the air would be a cheap alternative to a land-based strategy.

The Canadian manpower crisis of the first global war of the twentieth century was to have a seminal effect on all government plans for the conduct of the second. Memories of bitterness and division over compulsory military service meant that, for Mackenzie King, the "conscription question entirely dominated his thinking on questions of defence."[14] It shaped his government's policy as war approached, and it controlled it until war's end. The prime minister's course was cautious and aimed to avoid stirring up the embers of a divisive issue that could split the country. The key to his designs was the avoidance of a large-scale commitment of forces that could lead to heavy casualties on foreign fields, thereby provoking some form of compulsory overseas military service.[15] Mackenzie King's strategy of avoiding large overseas commitments involved proclaiming his government's intention to limit its liability, in both men and wealth, in the European conflict. When Britain asked Canada to provide large numbers of trained aircrew (through the BCATP) as the latter's principal contribution to the war, the request "must have seemed the answer to any Canadian politician's prayer." It then looked as though most of Canada's war effort could be concentrated on instructional duties at home. The BCATP would, it was hoped, provide a valuable contribution to the mother country without the prospect of insupportable Canadian losses. Even better, or so it seemed in 1939, Britain agreed to finance any Canadian air forces serving overseas.

The fall of France in June 1940 ended Canada's policy of limited liability. With the British Empire standing alone against the "seemingly irresistible Third Reich," Canada became Britain's senior partner in the war, and prepared itself to play a major role in the struggle.[16] The National Resources Mobilization Act, introduced the same month France fell, was the first serious step toward marshalling Canada's national potential. It required "persons to place themselves, their services, and their property at the service of His Majesty," but limited conscripts' service to Canada or her territorial waters.[17] All persons over sixteen years of age were obliged to register with the government. Eight million did, and of this number, 802,458 single men and childless widowers twenty-one to forty-five years of age were liable for military service beginning in August 1940.[18] In October 1941 a National Selective Service System was established in an attempt to regulate labour throughout the Dominion. It eschewed compulsory worker allocation, but imposed certain restrictions on employment to encourage workers to move to critical areas of the economy. The system also acted "largely through indirect means, and moral compulsions," to provide "the bulk of the manpower required by the Army, and a considerable proportion of that of the other services during the build-up period" from 1941 to 1943.[19] However, Canada's war effort was never total, as the Cabinet decided to allow many men to remain in non-essential industries as a way of avoiding an economic situation that would lead to a scarcity of consumer goods and widespread rationing.[20]

This policy was to cause manpower shortages throughout the country, and create a particularly difficult predicament for the army.[21] Even before the war began, the Chiefs of Staff of the three services made plans for a large ground-based expeditionary force in anticipation of, in their words,"'an immediate and overwhelming demand for active intervention'" on the side of Britain when hostilities commenced. The 1st Canadian Division was the first element of a force that was to grow to five divisions, two brigades, and attendant support troops.[22] The 1st Canadian Division was dispatched to

Britain at the end of 1939, but the army's "static role" early in the war disappointed many at home, and there was considerable pressure for a more active part for Canadian troops. Mackenzie King at first objected to taking "'the lives of any men for spectacular purposes,'" but in late 1941 he acquiesced on the grounds that public opinion and Canada's postwar standing in the world demanded a more tangible contribution to victory. This decision, and even the losses suffered by Canadian troops at Hong Kong and Dieppe, were, however, to have no immediate effect on the army's manpower situation.[23]

The first conscription crisis, in 1942, came about largely as a result of Japan's entry into the war. Public alarm about home defence was used as an excuse by Mackenzie King to significantly increase the land forces in British Columbia; by enlarging the home defence establishment, Mackenzie King hoped to avoid sending more men overseas. The expansion, however, caused gaps in the army's order of battle, and overseas conscription was proposed as a way to fill them. But the plebiscite of 27 April 1942, in which the government asked the Canadian people to release it from its commitment not to send conscripts overseas, was largely a political ploy by Mackenzie King to avert a split in his party, and in the country at large, between the pro– and anti–overseas conscription forces. By seeming to give something to both sides, Mackenzie King intended to avoid overseas conscription "until compelled to adopt it."[24] And, as the war progressed, it looked as if his wishes would be fulfilled.

In 1943 the Canadian war effort was "almost at its peak," and by the end of the year the RCAF was beginning the task of winding down the BCATP. In October 1944, with the "war virtually over and won," Mackenzie King was therefore "almost beside himself" when he learned that "an army of half a million men ... could not find 15,000 infantry reinforcements" to replace losses in the European theatre.[25] This was the beginning of the "second conscription crisis," which split the Cabinet and caused the resignation of the two key defence ministers, Power and Ralston.[26] Compulsory overseas service was finally imposed in late November 1944, when the government authorized 16,000 conscripts to be sent overseas. About 3,000 had reached their units by V-E Day.[27]

Whether overseas conscription was necessary from a military point of view is still open to debate. Stacey has claimed that compulsion convinced others to volunteer for overseas duty, and provided the needed margin of reinforcements.[28] E.L.M. Burns has been somewhat non-committal, but he appeared to doubt the value of the decision.[29] According to Granatstein, army casualty estimates were inflated, and conscription may not have been necessary to sustain the service.[30] More recent scholarship has examined the question through the lens of civil-military relations, and has reached harsh conclusions regarding the army's ambitions. R.J. Walker has portrayed the army and its minister J.L. Ralston as pressing for conscription not to win the war but, as Ralston admitted, "'to maintain the Canadian army.'" The "revolt of the generals" during the 1944 conscription crisis (when a number of senior army officers appeared ready to resign) has been described by Walker as "a massive intrusion by the army into the Canadian political arena."[31] Whatever the details of the situation, all commentators have agreed that the decision to introduce compulsory overseas service was primarily a political act. Walker has stated that it was Mackenzie King's clever way of thwarting the army's ambitions to defeat his government.[32] Stacey has argued that it was a rare example of one of King's eccentricities, in this case his "irrational intuition" that there was a conspiracy against him, becoming the basis for a "great act of policy."[33] According to Stacey, the prime minister's chief motive for introducing conscription was to appease the strong pro-conscription forces in the country and in his party and Cabinet, thereby preventing the possible disintegration of all three.[34]

But the influence of political circumstances should not divert our attention from the fact that military pressures precipitated the government's actions. Both the air force and the army demanded more manpower than was available. There were two major reasons for the RCAF's seemingly insatiable appetite for high-quality recruits: the unchecked expansion, until late 1943, of the BCATP, and high aircrew casualty rates, especially in Bomber Command. As for the land campaign, Stacey has written that one of the government's "two fundamental errors of military policy" during the war was its dispatch of a Canadian army corps to Italy in 1943. This not only diminished national control over ground forces, but with the army divided between the Italian and northwest European theatres, necessitated a high ratio of administrative to fighting troops.[35] The situation was further aggravated by the creation of two types of army, a conscript force for home defence and a volunteer force for overseas service. The home defence army was in a particularly invidious position, as it was "'neither in the war nor out of it,'" and was not involved in productive labour.[36]

However, it is clear that whatever errors may have been embedded in these decisions, the army must accept much of the responsibility for the situation in which it found itself. The army's insistence on a large force sustained by conscription led it to press the government for expansion and to lobby at every opportunity for compulsory military service to bolster its numbers.[37] But like its British counterpart, the army committed its most serious manpower errors when it inaccurately estimated wastage rates among the various combat arms. Terraine has attributed the British Army's inability to calculate combat losses to its "widespread misinterpretation of the nature of war, and a consequently flawed manpower policy."

The same estimation errors were duplicated in Canada when, as the war drew to a close, declining enlistments, "heavy casualties and a miscalculation of probable losses as between different arms of the service ... produce[d] a serious manpower crisis" in the army.[38] The miscalculation resulted from the army's inability to predict the disproportionate number of casualties suffered by the infantry in relation to other arms. The infantry lost ten times the numbers of the armoured forces, the arm suffering the next-highest casualty rate (76 per cent versus seven per cent), yet reinforcements were not allocated accordingly.[39] In addition, infantry constituted a relatively small portion of Canadian forces in Europe. Of an overseas contingent of about 250,000 in late 1944, only some 34,000 were in the infantry – the highest "tail to teeth" ratio among the major combatants.[40] As Granatstein has said, "The fault ... was not a shortage of men, but a lack of flexibility in their management and use."[41] This was the real root of the Canadian conscription crisis of 1944.[42] Even though the trend toward increased wastage rates in the infantry was known as early as 1943, "the lack of a timely decision as to how large an army we were going to mobilize" created "many of the inefficiencies" that were later experienced in manpower use.[43] These inefficiencies, among other things, led to a misallocation of manpower within the army, and, some have claimed, among the army, navy, and air force.

A superficial reason often given for Canada's World War II manpower crises is that the air force jealously hoarded highly qualified volunteers for an "over-expanded aircrew training program."[44] It is true that by October 1944 the RAF had accumulated a surplus of 60,000 qualified aircrew,[45] and that the RCAF had "10,500 trained aircrew surplus to all needs,"[46] but the crises had other causes. The surpluses resulted from decisions made in 1941 and 1942, and of course there was no way during those years to precisely predict 1944 requirements. When the manpower allocations were made, the air force had priority both in Allied and in Canadian plans, and the "RCAF's manpower needs for the period 1 Jan 43–31 Mar 44 were well within the estimate which had been submitted to the Cabinet War Committee during September 1942."[47] While the aircrew surplus has been noted by many scholars, a lesser-known fact is the RCAF's use of 80,000 "'A' category men" (those who were "physically fit for overseas service") for ground duty. Some of these men were required for overseas service, including 6,000 radar technicians, but the vast majority stayed in Canada.[48] Even as army manpower sources were drying up in May 1943, the RCAF started enlistment proceedings for another 6,000 "A" category ground crew. The problem of the air force employing large numbers of "A" category men in Canada was not rectified, despite lengthy negotiations, until September 1943, when an understanding between the army and the RCAF was reached, and the RAF agreed that it would no longer enlist ground crew whose medical category made them suitable for army service overseas.[49] In retrospect it seems surprising that it took so long for the army and the air force to coordinate their recruiting activities, but this was just one example of Canada's disjointed human-resource management system.

During the first two years of war there was no trouble finding manpower for most military requirements, because as much as one-third of Canada's non-agricultural workforce of three million was unemployed, and recruiting quotas were modest.[50] However, by mid-1941 the labour surplus had disappeared, and factory pay became a big incentive for many to remain in civilian life rather than join the armed services.[51] In the absence of a national economic policy there was, according to a government report, "'competition between the armed forces and industry, between war and non-war industries, and among industrial concerns generally'" for scarce labour resources.[52] In this competitive atmosphere the RCAF was dealt a very strong hand and played it well. Its trumps, especially for aircrew, were glamour, travel, new skills, high rates of pay, and a good chance of a commission.[53] Despite the inefficiencies caused by this labour-allocation system, or lack of a system, a mutually acceptable means of distributing human resources among the fighting services and industry was never satisfactorily determined. This had far-reaching repercussions. For war production, it meant, after 1943, a reduced output of base metals and coal, which adversely affected the munitions supply.[54] Not just industry felt the pinch, however. Even though the army had experienced manpower shortages before, it could not really claim serious hardship until the middle of 1943. The Canadian Army began to suffer significant battle casualties in July 1943, but the "manpower accessible through voluntary methods had been largely exhausted" six months previously.[55] By the second half of 1944 the army was desperate for manpower and hoped to obtain it from other sources.[56] The air force was in a position to offer only the most limited kind of help. In late 1943, for example, some RCAF aircrew candidates who had not started flying training were directed to the army, but what the land forces really needed in 1944 were properly trained and "efficient" reinforcements, not physically fit raw recruits.[57] For better or for worse, decisions taken early in the war determined Canada's manpower posture toward the close of the conflict.

This is not to say that the air force was blameless. The RCAF's main difficulty was that its manpower requirements were being driven by British priorities. The RAF had been fooled once, during the "phoney war," into underestimating wastage rates, only to find itself short of necessary manpower. This experience influenced its future plans to keep a cushion of 7,000 to 8,000 pilots awaiting precious advanced-training spots. As events transpired, wastage was less than anticipated, but by October 1942 mobilization in the United Kingdom had gone further than it had by the end of World War I, with 30 per cent of the national population in the forces in 1942 compared to 29 per cent in 1918; and yet the services required still more men.[58] Changing strategic, tactical, technical, and organizational circumstances forced the RAF to continually revise its manpower estimates. On the one hand, it is hard to disagree with the RAF view that in mid-1943 it was difficult to predict wastage rates for 1944 and 1945 "air battles which could only be dimly foreseen."[59] On the other hand, sudden changes in British aircrew requirements certainly gave Britain's allies the impression that the Air Ministry "had only the vaguest idea how many aircrew they actually had on hand, how many were in the training stream, or how many were needed."[60] The RAF's rejoinder was that when surpluses were acknowledged, it was the only service urged to reduce its training organization while at the same time maintaining front-line units at full strength.[61] For the Air Ministry staff administering the air-training scheme, the organization they directed and the huge force it produced may have existed in pre-war film fantasies and novels, but the procedures required to manage such an immense undertaking had never been devised, let alone put into practice. When the taps on the training pipeline were opened to maximum, only a few drops came out at first. The stream later turned into a torrent, but turning the taps off at the BCATP source did not stop the flow reaching Britain for some time. In any event, the air force believed that more was better, and given the competitive nature of the manpower game, there certainly was no incentive to turn the taps off too soon. In the end, a much greater number of aircrew were trained than could have been imagined in 1939; but sheer numbers were not the only criterion of success.

E.L.M. Burns has concluded that, because the supply of personnel was finite, the most important lesson learned in World War II was the necessity for economy in the employment of manpower. Although his work has focused on wastage in the Canadian Army, he also has speculated that the RCAF, and the Royal Canadian Navy, could have used their manpower "more economically," and would have endured even greater "inefficiencies" had they been dealing with larger numbers of men.[62] One way to evaluate this criticism is to examine how efficiently the RCAF used the manpower it recruited, by looking at the success rate of the BCATP. Of the 91,166 RCAF aircrew who commenced training under the plan, approximately 80 per cent graduated in one of the aircrew categories. Of those who failed one type of aircrew training, "better than 50 percent" were salvaged and re-entered the training stream in another category. Perhaps the most impressive evidence of the plan's efficiency was that almost 93 per cent of entrants were eventually employed by the armed services, either receiving their wings or remustering to another trade in the air force or to duty in the army or the navy.[63]

However, certain factors under RCAF control did lead to wastage. Extremely stringent medical standards, and high failure rates in certain courses (for example a rate of over 40 per cent on some pilot intakes), led to the dissipation of scarce human resources. No precise figures on efficiency are readily available, but one way of illustrating the magnitude of the problem would be to examine the fate of a hypothetical group of 10,000 RCAF applicants in 1941–43, the years when four out of five applicants enlisted.[64] To be considered for aircrew duties our applicants had to be male, young, and British subjects "of pure European descent." They also had to be motivated to fly, and have the necessary physical, educational, and intellectual attributes. However, not every applicant wanted to become aircrew. Some who joined the RCAF had no desire to fly, while others had skills that led them to be sent to the ground trades where they were most needed. Some, of course, were rejected out of hand because of obvious deficiencies.

The majority of the men who were enrolled as aircrew trainees were from seventeen to twenty-two years of age, and, unlike the army's enlistees, almost all had been students or gainfully employed before joining. Just over 40 per cent left manufacturing or clerical jobs to enter the RCAF, and 8 per cent had worked in agriculture.[65] Over 20 per cent of aircrew candidates were rejected because they failed the enlistment physical exam,[66] and most of the rest were eliminated owing to poor results on the recruiting officers' assessment of character, deficiencies in their formal education, or a finding of low learning potential on the selection tests. For all of these reasons, of our hypothetical 10,000 who "approached"[67] a recruiting centre, only 6.5 per cent, or 650, would have been enrolled in aircrew training.[68]

Those who began aircrew training under the BCATP stood an 83 per cent chance of graduating.[69] Of our 650 hypothetical aviation candidates, 536 would thus have earned their wings in one of the aircrew categories. What the 83 per cent figure hides, however, is the number of aircrew who were selected for one kind of duty, usually that of pilot, but did not complete their training in that area. Even though they might eventually graduate from the BCATP, such men had to be held back, reselected (in most cases), remustered to another category, and retrained. The RCAF considered that it suffered from high rates of training wastage and reselection,[70] but, as we have seen, it was unable to reduce those rates significantly before the BCATP was wound down.

Of our group of 536 who would have earned their wings, about two-thirds, or 338, would have been posted overseas,[71] and about 60 per cent, or 203, would have ended up serving with the RAF.[72] Until Canada established its own administrative system, some of this precious cargo would have been lost, as we have seen, to wasteful RAF training and personnel policies. Bomber Command claimed the largest share of Canadian airmen sent to Britain, so 40 per cent of the 338, or 135, could have expected a posting to that Command in 1944.[73] About 17 per cent of this number would have been eliminated in advanced training in Britain, leaving 112 to reach operational bomber squadrons.[74] Of this group, one typically would have been classified LMF, six diagnosed as suffering from a neurosis, eighteen wounded, and fifty-four (just under one-half) killed or reported missing on operations, leaving thirty-three to survive the war unscathed.[75]

The foregoing hypothetical case study illustrates the RCAF's chief manpower difficulty – a large number of recruits were required to procure one trained aviator. About ten men had to come through the recruiting-office door for every trained RCAF aviator delivered by the BCATP.[76] Moreover, the applicants who were unsuited for flying duties were not necessarily redirected by the RCAF to where they were most needed. The RCAF estimated that each combat aviator required five people in uniform to support him,[77] and in the case of men chosen for ground-support jobs, it picked the best of its non-aircrew servicemen to fill those positions. Of our representative group of 10,000 candidates, about 3,600 would have joined the air force in some capacity; this would have left almost two-thirds available for employment outside the RCAF. In real terms, this translated into some 380,000 men who were rejected by the RCAF from October 1939 to June 1944.[78] Even though they were available to be redirected into essential wartime occupations, including the army, many may have been able to avoid compulsory reassignment.

There is evidence to suggest that "in some sense each service draws manpower from a unique constituency,"[79] and that, given the competition for manpower in Canada during World War II, it would not have been an easy matter to reallocate RCAF personnel to the army, even with better forecasting and planning. Unlike the army, which often "took men who could not get into the other services and who failed to find a place in industry,"[80] the air force selected recruits who generally were well-educated or possessed technical skills. Men with marketable skills who were rejected by the RCAF would not necessarily have found service with the army appealing. World War I had taught people that "soldiering was a filthy business ... demanding enormous casualties,"[81] and many who were not accepted by the air force may have been able to find civilian or service occupations that spared them from front-line combat.[82] But even if the army did acquire them, there is no reason to assume that, before 1944, it would not have used tradesmen in one of its *armes savantes*, as the French army referred to its technical corps,[83] rather than in the infantry where the need would later exist. Terraine refers to the British Army policies that placed the infantry's best men in the technical arms and in élite forces such as the Special Air Service as "vicious" and "calculated to wreck the infantry."[84] In John English's words, the infantry became the "legitimate dumping ground for the lowest forms of military life."[85] As for those with higher qualifications, it is revealing that even though the Canadian Army was to develop a shortage of other-rank infantry replacements, from 1943 on it had a surplus of officers from which it was able to provide the British Army with over 600 volunteers under the CANLOAN scheme, the vast majority as infantry subalterns.[86] This was another example of people being permitted, and in this case encouraged, to make choices that did not necessarily coincide with the national good as defined by the state. But even in total war, which World War II was not by many measures, most manpower-allocation objectives could not be achieved by coercion. Complex bureaucratic structures and the need to maintain public support resulted in many loopholes that the intelligent or those skilled in a trade could use to avoid reassignment under duress.

This made securing the necessary quantity and quality of human material a difficult undertaking. In this respect the RCAF and the Canadian Army shared a number of problems. Both services required over one year to prepare men for combat,[87] imposing a long lead time on planners. The two services were also required to continually modify their training as the changing nature of the war and advances in technology forced them to restructure their military manpower priorities. Numerous training delays, and problems reassigning men to different occupations, ensued.

The main difference between the army and the RCAF was that the junior service unquestionably attracted the best-educated recruits, and, of the recruits whose formal education was meagre, the ones with the best learning potential. The rigorous air force selection system also ensured that the RCAF retained the most desirable candidates, when the army was forced to accept many recruits of doubtful value.

The RCAF's greatest failure in the manpower field was its high wastage rate in pilot training due to its use, early in the war, of flawed selection criteria, and to deficiencies in the training system. This resulted in a great deal of additional effort and delay in reselecting and retraining aircrew. From the air force's point of view, however, this was not a fatal error.

The RCAF's untrammelled expansion can be explained by the air force's attractiveness to Canadian youth and by the high priority the government assigned to the air effort. In the final analysis, it was the government's responsibility to provide clear instructions to industry and to the army, navy, and air force concerning manpower allocation. Its inability to offer unambiguous direction allowed a competition for manpower to develop among the four sectors, a competition that the air force won, or at least tied for first place with industry. The loser was the army.

It has been suggested that Canada did not have enough human resources in World War II to meet its military and civilian requirements,[88] but the point is moot, for it was the government's obligation to balance national commitments against available resources, and in this respect the government failed, or at least allowed a decision by default.

Even if the government did not make the most efficient use of its manpower, we should recognize that Canada made an extraordinary contribution, for its size, to the Allied war effort. The government registered every worker in the country, and raised very considerable military forces, putting one in eleven Canadians, and almost one-half the eligible male population, into uniform.[89] It will be recalled, however, that the government's intention was to minimize casualties as far as possible.

The thought of the cream of the nation's crop being thrown into the maw of another ground war was enough to terrify most politicians.

As we have seen, the importance attached to the Allied air effort, especially the strategic bombing campaign by both the Canadian and the British governments was intended to avoid casualties of World War I magnitude. But 55,573 RAF aircrew, many of whom were Canadian, were killed in World War II. This number can be compared to the 38,834 officers of the British Empire forces who lost their lives in World War I. As Terraine has put it, the

slaughter of the nation's élite [in World War I] was widely regarded as the most tragic and damaging aspect of that war ... Yet ... by and large RAF aircrew were exactly the same type of men as the officers of 1914–18; it is salutary to see how the pursuit of a "cheaper" policy brought in its train only a much higher cost.[90]

The Canadian situation was similar. During World War II, 17,101 Canadians were killed while serving outside of Canada with the RCAF or the RAF, compared to 17,683 Canadian Army "battle casualties."[91] This was the antithesis of what Canada's political leaders had envisioned in their plans.

8 Conclusion

Forty years and two world wars after the first powered aircraft staggered into the sky, aviation had changed beyond recognition. By 1944 huge fleets of multi-engined aircraft, capable of flying in almost any weather, could drop tons of explosives on targets hundreds of miles from their bases. But more than the aircraft had changed. Large numbers of aircrew with special abilities were now required to man the new machines, and training times had increased from weeks to months and even years. Yet despite these differences, the crews of the most modern combat aircraft were still human, and displayed many of the same frailties as the earliest aeronauts. Four decades after their effects had first been studied, cold, lack of oxygen, and psychological stress continued to hinder the crew member from effectively carrying out his duties. While great strides had been made by the end of World War II in overcoming the physical limitations experienced by aircrew, psychological considerations presented a much greater challenge. Of all the human aspects of flying investigated by the experts, the mental make-up of the aviator proved to be the most resistant to explanation.

As an organization responsible for tens of thousands of aircrew from 1939 to 1945, the RCAF developed a keen interest in psychological issues. This interest was reflected in the RCAF's selection and training policies, and in its policies governing psychological casualties. The RCAF's experiences in developing these policies, as well as the history of its responses to Canadian manpower problems in World War II, highlight some aspects of the "managerial metamorphosis" that took place in the conduct of air warfare during the first half of this century. Technocratic élites, especially Canadian psychologists and British physicians, contributed to the raising of Commonwealth air forces on an unprecedented scale. In so doing, they effected profound changes in the air forces as well as in their own professions.

Manpower, one of the three "ultimate regulators" of a nation's war effort, is a unique variable in the resource equation.[1] Unlike the other two variables, food and fuel, which have artificial substitutes, there is no replacement for skilled personnel. The long lead times required to produce aircrew and other highly qualified combatants leave little room for error when limited national resources are allocated to expensive training programs. Modern states require the assistance of specialists in many areas to manage their resources efficiently.

After the political authorities lay down general directions, the implementation of human-resource strategies normally rests with the administrative apparatus of the state. To effectively execute the military aspects of such strategies, the "brain" of the armed forces, composed of headquarters and other staffs, requires expert knowledge which often must be drawn from civilian sources. From whom this information is acquired often determines the nature of the advice given. As the saying goes, "to a man with a hammer every solution is a nail." The Canadian experience of raising a great air force during World War II clearly illustrates this truism.

The RCAF's primary aim in recruiting potential aircrew was to select those who had the best chance of completing training. The selection criteria developed by the medical profession and by the Canadian psychological community focused, at the entry stage, on physical health (especially good vision), youth, and learning ability. Next, tests to measure aptitude for acquiring complex skills, such as those required to pilot an airplane, were given. The final part of the selection process involved actual training where, as students confronted more and more difficult challenges, attrition became increasingly costly to the air force.

The question of which professional group should act as the arbiter of truth in aviator selection is an issue that remains relevant today. Psychologists and psychiatrists continue to debate, with each other and even among themselves, whose methodology is the most appropriate for finding people who have "the right stuff" to become military and civilian pilots. In Europe this controversy has taken on a degree of urgency, as the continuing process of political unification within the European Union has meant that its member states must adopt common standards for aviation personnel. The struggle for ascendency between and within professional communities in the aircrew-selection arena has only just begun.[2]

One of the questions this study has sought to answer is whether RCAF aircrew constituted the cream of Canada's manpower crop. The answer probably depends on individual value judgments and on the social paradigm that one endorses, but in the eyes of Canada's leaders and of the Canadian public in World War II there was little question that the air force's finished product was the best the nation had to offer. Today, with a Canadian air force that includes both female and male aircrew, many who serve in the Canadian Forces believe that the same assumption is true.[3] The supposition which underlies this belief is that young, healthy people who have demonstrated the ability to master complex tasks, to adapt to changing situations, and to make decisions under extreme pressure, possess the essential qualities of an aviator, and, by extension, as "Chubby" Power suggested in 1941, vital qualities for leadership in our society.

Another aim of this book has been to evaluate the RCAF's effectiveness in shaping and using its valuable manpower resources. There were two ways for Canada to affect how the nation's aircrew were trained and employed: first, by formulating its own training and employment policies, and second, by influencing Allied war-fighting strategies. Canada achieved better results in the first area than in the second, because selection and training, two of the three main determinants of efficiency in aircrew use, were, as we have seen, largely under RCAF control.

Canadian doctors had been involved in aircrew selection since World War I, particularly in association with RFC/RAF Canada, but drastic economies between the wars wiped out the infant medical branch of the Canadian air force. The CAF/RCAF's dependence on army physicians for the twenty years after 1918 left a large void in Canadian aviation medicine. This was evident when some outdated medical standards from World War I were used to screen air force recruits at the beginning of World War II. When a more effective RCAF aircrew-selection process was finally developed, it was chiefly Canadian psychologists who created that unique system. Their efforts produced a pragmatic and effective procedure based on written and psychomotor tests of aptitude and learning ability, but it did not reach the peak of its efficiency until the bulk of BCATP aircrew had been processed. Although Canadian psychologists in developing the World War II RCAF aircrew-selection system adopted criteria that forced complex human characteristics into simplified categories, they believed there was little alternative given the demands of the war. In the end, they were satisfied that they had found procedures which furnished scores that correlated well with performance in training, and these procedures were adopted by the air force to replace subjective judgments in aircrew selection.

Unlike the selection process, RCAF training remained in the hands of professional officers during World War II. It did not benefit from outside expert advice to the same degree as the RAF system, which called on civilian Canadian psychologists for advice in training methods. The RCAF's neglect of expert advice need not have had serious consequences because, as early as 1918, Canadians had modified the British Gosport method to create the Armour Heights System, their own very effective method of training pilots. Yet if anyone in 1939 remembered how it worked, he did not bring it to the attention of the air force authorities. The rapid wartime expansion ushered in by the BCATP overwhelmed the small RCAF training group, and the many new flying schools turned out flawed products for most of the war. An efficient training system, with many of the same principles as those found at Armour Heights a generation before, was implemented only after most trainees had passed through the BCATP. The resultant wastage at advanced stages of training was particularly serious, because the farther a trainee without sufficient aptitude was allowed to progress before being failed, the greater was the cost to the air force. Even worse, variations in standards from school to school probably meant that a number of men who would have made useful pilots were failed. In retrospect, it seems that RCAF staff officers did not take the trouble to review their service's past and to see what had been learned in its early years. If an efficient training program had been initiated in 1939 or 1940, and if a sufficient number of experienced RCAF instructors had been assigned to oversee its creation, an aircrew-training system that minimized wastage would have been in place sooner.

Whether Canadian aircrew could have been employed more effectively in World War II to reduce combat losses is a more difficult question to answer, as many of the decisions that had an impact on RCAF personnel policies were made by the British air force. However, the RCAF was able to salvage proportionately more manpower than the RAF once it established an independent administrative system overseas, which allowed it to implement its own personnel policies, rather than having to follow those of the RAF. Although in terms of total numbers the savings were small, the creation of policies aimed at the productive employment overseas of as many aircrew as possible paid large dividends, as each trained flyer in the United Kingdom was literally worth more than one-half his weight in gold.[4] Moreover, as we have seen, salvaging one operational aviator overseas could make up to ten men in Canada available for other employment. A further important aspect of aircrew-personnel policies was that they afforded Canada control of its citizens and allowed RCAF authorities to treat air force personnel in ways that conformed to national social norms. In this respect, these policies can be viewed as a significant assertion of Canada's sovereignty.

In 1939 plans for the prevention and treatment of air force psychological battle casualties held out the promise that a great deal of aircrew manpower could be conserved. The problem of psychological losses, and the basics of treatment, had been identified during World War I. Yet between the wars the problem was largely ignored by air forces, because psychological losses were seen as something that could be prevented by appropriate selection methods, good training, and inspired leadership. In the face of heavy casualties in World War II, however, these measures could not always be fully implemented, and "flying stress" became a serious concern. The British approach to the problem was varied and drew on the advice of experts in neuropsychiatry, who proposed, among other things, better living conditions for aircrew, reduced operational stress, and psychotherapy for the afflicted. The most controversial of Britain's policies on flying stress was the one devised for aircrew who had been identified as lacking in moral fibre. At first, the service view that exemplary disciplinary action against a minority would be sufficient to maintain morale prevailed. However, it was ultimately realized, after considerable prodding from the Dominions, that Britain's "waverer" policy was being applied to cases for which it had never been intended, and that this was wasting manpower and inflicting injustices on aviators who did not deserve disciplinary treatment. LMF procedures were then modified in an attempt to correct their deficiencies, but they continued to emphasize bureaucratic convenience and operational necessity at the expense of justice in individual cases. Once again, the RCAF was forced to rely on British procedures for some time until it was able to implement its own approach. By 1944 the RCAF system for handling LMF cases was finally in place, and Canadian standards, based largely on legal and political considerations, were applied to airmen who had been suspended from flying duties and sent to RCAF reselection boards. However, RAF procedures continued to govern Canadian operational squadrons, especially the squadrons in Bomber Command.

Perhaps the most important lesson about personnel management learned by the RCAF in World War II was the importance of conserving manpower at all stages of aircrew selection, training, and employment. Effective selection tests were needed to give the men who proceeded on to training a better chance of success, and to avoid overburdening instructors with candidates whose chances were poor. In flying training itself, more standardization was required to improve the calibre of instruction, to reduce wastage, and to produce better-qualified students. On operations, it was necessary to limit casualties among experienced aircrew, especially casualties arising from causes over which air force leaders could exercise some control.

While much of this might seem like "common sense," this book has shown, if anything, that such notions were certainly not common, and sometimes not viewed as sensible. This was due in some measure to varying opinions on what constituted a sensible course of action. For example, with respect to selection methods, psychologists and psychiatrists held very different views. With respect to training, Tedder's "commonsense" approach of 1934–36, which was similar to the World War I Gosport method, was declared "impracticable" by the RAF in 1936. And the "commonsense" approach of RFC and RNAS doctors to psychological casualties in World War I was, in the next global conflict, considered operationally unfeasible by Bomber Command and ineffective by some medical specialists.

In establishing policies and practices for human-resource management on a national scale, choosing which kinds of experts will define the problems to be faced and the forms of knowledge to be employed in overcoming those problems can be crucial to the success of the endeavour. Psychologists, using statistical methods of inquiry for the most part, shaped much of the RCAF's, and the RAF's, selection processes. But the treatment of psychological casualties – in fact all psychological treatment in both air forces – remained the prerogative of the medical profession, and outsiders were excluded.[5] Psychologists and physicians each had their own conceptual models to explain human behaviour, and brought differing skills and prejudices to the problems at hand. In addition, each group saw the war as a way of advancing its prestige, and did not hesitate to promote its particular skills as the most efficacious for solving wartime dilemmas.

At the start of the war Canadian psychologists were ill-prepared as a profession to make a major contribution to the nation's war effort. However, by adopting a strategy designed to advance the cause of psychology during the war, and by having many of the most eminent civilian researchers in the country carry it out while in uniform, the psychologists were able to elevate their specialty to a position of prominence in the scientific community. By focusing on an area of immediate interest to the government and to the armed forces – aircrew selection – they attracted substantial funding for their work and were able to shape certain aspects of air force policy.

British psychiatrists made similar gains on their side of the Atlantic, but their task was easier in many respects because of the prestige attached to the medical profession in Western society. Some of the most celebrated physicians specializing in the field of human behaviour worked for the RAF during World War II, and they had a significant influence on policies governing aircrew who, for psychological reasons, could not or would not fight.

Both psychology and psychiatry had an important effect on the management of Canada's human resources, the former by selecting for military service those whom its practitioners considered "the best" among the mass of the populace, and the latter by helping to conserve some of the cream of the crop. Despite their failings, both disciplines helped Canada to field armed forces that were huge in proportion to its population.

However, most senior professional military officers knew little about psychology and psychiatry, which hampered their ability to understand the biases of their specialist advisers, and consequently to comprehend how such biases might affect the nature of the advice they received. We might also ask how often those responsible for directing the armed forces were guided by past experience. Martin van Creveld has suggested that the specialization necessitated by sophisticated technology "may contribute to a loss of institutional memory."[6] In this context, perhaps the nature of Canadian air force institutional memory from 1918 to 1945 is worth discussing briefly. World War I afforded interested observers a great deal of information about air warfare. Abundant experience from that conflict survived in individual memories, and filled books, official reports, and journals. However, since little was done to sift the relevant material, to organize it for specialized study, or to make it accessible, most of the war's intellectual bounty was of little immediate use. The structure of the RCAF between the wars aggravated the problem. With few resources invested in headquarters staff or in administrative training, almost no one was assigned to systematically evaluate the lessons of World War I. And even when some effort was directed toward this task, it appears that it was not always accomplished satisfactorily. Sir Andrew Macphail's official history of the Canadian medical services, for example, has been described by Desmond Morton as "thin and wholly inadequate."[7] If the chore of absorbing the hard-won lessons of the war had been perceived as more urgent, perhaps fewer of Canada's cream of the crop would have been squandered in World War II.

The other means of reducing RCAF wartime losses were more problematic. Whether Canada could have significantly influenced

British and Allied strategy to reduce aircrew wastage, especially in Bomber Command, is a much more difficult question to answer. The Mackenzie King government was rarely consulted on matters of strategic importance, and, even if it had been, there is no reason to suppose that its advice would have been accepted by Arthur Harris any more than the British government's was. Spencer Dunmore and William Carter have been extremely critical of the role of Charles Portal, the RAF's CAS, in the strategic bombing offensive against Germany, because he failed to compel Harris to adhere to the policy objectives established by Harris's superiors, especially as the war drew to a close.[8] In an attempt to prove he could win the war from the air, Harris insisted, almost to the end, on staging mass raids, with heavy losses to his forces.

In the matter of Canadian independence, Carter and Dunmore have been equally critical of the Mackenzie King government's policy of trading sovereignty for money. "Gus" Edwards, the senior RCAF officer in Britain in October 1942, drew attention to this situation when he told officials in Ottawa that problems with Canadianization were largely caused by Canada's not paying for the upkeep of its air forces overseas.[9] Once Canada agreed to pay its share beginning in early 1943, issues hinging on sovereignty were settled to the Mackenzie King government's liking much more easily.[10]

The question still remains, however, whether Canada could have affected the course of events to reduce its losses, particularly in Bomber Command. As the supplier of most of the Command's trained aircrew, many of whom were its own citizens, Canada possessed a powerful lever to influence RAF policy. But even as casualties mounted at an alarming rate, the Mackenzie King government continued its full support for Harris's campaign by sending over a steady supply of Canadian "bodies," even when the campaign's effectiveness was uncertain.

If Bomber Command's main effect on Germany was psychological, as recent evidence has suggested,[11] then it did not matter from an operational point of view whether Harris conducted 1,000- or 500-bomber raids. A smaller force would have been adequate for maintaining psychological pressure on the German population, while also providing the necessary squadrons for the very effective precision attacks that were carried out later in the war. Restricting the size of the strategic bomber force, especially in the early stages of its growth, would have preserved manpower in several ways. Smaller numbers of aircrew would have allowed higher-quality candidates to be chosen, and longer training periods would have reduced casualties. In addition, fewer aircrew in the training system would have reduced the bottleneck effect for trainees passing through the bomber OTUs, and operational squadrons would have been required to send fewer experienced flyers to form the OTU training cadres.[12] Finally, less-intense operations would have reduced losses of all types, including those due to flying stress, and would have permitted crews reaching the end of their psychological tethers to be rested in time to allow them to continue on operations for a longer period. Of course, if losses in the air had been reduced, more manpower could have been made available for the other urgent demands of the war.

Perhaps it was too much to ask of a junior partner in the alliance against the Axis, particularly one led by a prime minister who accepted that he held a minor place among the leaders of the great nations, to intervene in matters of grand strategy. In retrospect, though, it seems ironic in light of the lessons of World War I that Canada was once again prepared to surrender its young men to the control of an empire ready to commit those young men to a costly bombing campaign of dubious value. The Dominion entered World War II intent on keeping its casualties to a minimum, but during the last half of the war it assented to a policy that decimated the cream of the crop.

From the perspective of national sovereignty, the lack of Canadian involvement after World War I in matters as important as aircrew selection and training compelled this country to trust in the knowledge other nations possessed, and to accept the cultural biases that accompanied that knowledge. The degree of Canadian dependence varied, however, with the efforts it was prepared to make on its own behalf. In the case of aviator selection, because most BCATP facilities were located in Canada, the RCAF was required to develop its own procedures, and it therefore maintained a good deal of independence in this area. A similar situation existed with respect to basic flying training; as we have seen, however, the outcome was different in certain ways. As for the psychological casualties of air combat, the lack of Canadian expertise and the great distance between Canada and the European theatre of operations forced RCAF authorities to rely on the judgment of others for most of the war.

The involvement of civilian technocratic élites in Canadian aircrew selection helped to institute a system that was not only more effective than that of World War I, but also more attuned to national sensibilities. Conversely, the absence of expert advice on methods of flying training resulted in a system that in many respects was worse than the one in place in 1918. Ironically, this caused unnecessary wastage at a time when Canadian psychologists were being employed by the British air force to improve its training methods. But the action that illustrated Canadian dependence on others most clearly was the RCAF's acquiescence in British policies regarding aircrew unfit for flying duties. Until Canadian political and legal élites imposed their own solutions, the RCAF condoned an aircrew-disposal system that was not only seen to be unfair, but was also known to waste valuable aircrew resources.

Given the economies imposed by peace on military forces, it was probably too much to expect that those officers of the interwar period who rose to high rank in the RCAF during World War II would have been adequately prepared by their interwar duties to deal with all the enormous problems that confronted them from 1939 to 1945. But minimal administrative and command experience among senior RCAF leaders played a role in creating the difficulties that contributed to the dissipation of Canada's manpower during World War II. It is true that Canadian air force leaders could have done better, but it is well to remember that the achievements of the RCAF during World War II far outweighed its shortcomings.

We have seen how civilian technocratic élites were mobilized to help Canadian society manage some of its most valuable human resources, even if the results were not always those anticipated by military and governmental authorities. The lesson here for the leaders of today and tomorrow is that it is just as important to be aware of the principles of human-resource management as it is to have operational and technical expertise. For unless those who direct our armed forces understand these principles well, Canada may once again be forced to accept the judgment of its allies as to how many of its citizens will die in battle, and in what ways.

APPENDIX A

Bomber Command Psychological Casualties

The first figures in the text that related to the calculation of the incidence of Bomber Command psychological casualties summarize "the occurrence of neurosis in Royal Air Force air crew." Symonds and Williams produced several statistical studies of psychological disorders among RAF aircrew, particularly FPRC Reports 412(g), 412(i), 412(L), and 547, which are listed in the bibliography and are the principal sources for the data presented in this appendix.

The numbers in Table A.1 are from the annual reports[1] on "psychological disorders" among RAF aircrew prepared by the service's neuropsychiatrists and summarized in FPRC 412(L). Most cases were in the categories "neurosis" and "lacking confidence." The categories "not having a neurosis but unsuitable for flying duties" and "considered normal and returned to duty" were first reported in 1944–45.

The next figure in the text provides an estimate of the total number of cases of psychological disorders dealt with by the RAF medical branch. No comprehensive study was done by Symonds and Williams; however FPRC 412(k) indicated that unit MOs treated about 70 per cent of the cases of aircrew who saw their MOs "for complaints which have no physical basis, or who having no physical disability state[d] that they [felt] unable to carry on with flying duties." For the 30 per cent of aircrew whose treatment was not attempted on station, 30 per cent (9 per cent of the total) were diagnosed LMF and dealt with by the executive branch, 60 per cent (18 per cent of the total population in the study) had a "bad or hopeless prognosis," and 38 per cent were "under training and considered unsuitable for aircrew duties." No explanation was given as to why the categorization of the 30 per cent of

Table A.1
"Cases [of RAF aircrew] referred for psychiatric opinion," 10 February 1942–9
February 1945

	1942–43	1943–44	1944–45	Totals
Neurosis	2,503	2,989	2,910	8,402
Lacking Confidence	416	307	306	1,029
Unsuitable for Flying			108	108
Normal			129	129
Totals	2,919	3,296	3,453	9,668

aircrew whose treatment was not attempted on station exceeded 100 per cent, but in other reports numbers in excess of 100 per cent resulted from individuals being assigned to more than one diagnostic category.

There is no indication of how many of the total number of cases first seen at the unit level was eventually referred to specialists; however, a conservative estimate might be 30 per cent. This allows for that portion of the 70 per cent who were initially treated by unit MOs but eventually referred to specialists, and those in the 27 per cent of the total number (above) who were in the LMF or "bad or hopeless prognosis" categories but were sent to neuropsychiatrists. Therefore I suggest that of all the cases of neurosis seen by the RAF doctors, about 30 per cent were examined by specialists on referral.

If one accepts that the figure of 9,668 aircrew seen by specialists or classified LMF represents 30 per cent of the cases of psychological disorder reported by the RAF medical branch, then the total number seen by all RAF doctors (specialist and non-specialist) or classified LMF over the three-year period would be 32,226, or approximately 30,000. (From this point forward all figures will be rounded off, because, as is explained in chapter 5, I am attempting to establish orders of magnitude, not precise numbers.) This yields an annual figure of about 10,000 aircrew who were identified by the RAF medial branch as suffering from a psychological disorder. I believe this to be a reasonable figure when one considers that the RAF maintained four hospitals to provide "intense and prolonged psychiatric treatment," as well as twelve NYDN Centres in the United Kingdom and two overseas. According to one postwar source, unit MOs were expected to treat only "the mildest cases," while the rest were to go to "the nearest NYDN Centre." The only statistics I have been able to find relating to hospital or NYDN Centre treatment are for RAF Hospital Matlock, which treated both aircrew and ground personnel, but not officers. Its annual admission rate was about 1,200.[2]

The next estimate required is the number of cases of "psychological disorder" that can be attributed to Bomber Command. Percentages vary from a low of 30 per cent (FPRC Report 412(L)) to a high of 65 per cent (FPRC Report 412(d)) depending on what definitions were used. The FPRC reports show an average from February 1942 to February 1945 of about 37 per cent,[3] which I have rounded off to 40 per cent. If multiplied by the estimate of 10,000 psychological disorder cases a year in the RAF, this yields 4,000 cases a year in Bomber Command.

Rather than established or actual strength, I have based my calculations of the incidence of psychological disorders in Bomber Command on the number of aircrew who were in or passed through the Command in a one-year period, because this represents the manpower pool within which the psychological casualties occurred. The total number of flying personnel in the Command for the calendar year 1943 was 79,254 (34,269 in the OTUs and 44,985 in operational units),[4] which I have rounded off to 80,000.

Dividing the annual psychological casualty rate of 4,000 by 80,000 gives an average incidence of psychological disorders in Bomber Command of about 5 per cent of all those who were in the Command in a given year.

The approximate casualty rate for Bomber Command is based on Webster and Frankland's figures on aircrew casualties, operational and non-operational, for 3 September 1942 to 2 September 1943 (19,096) and for 3 September 1943 to 2 September 1944 (25,896), the average of which is 22,496.[5] I have rounded this figure to 20,000.

APPENDIX B
Selected Statistics of the Special Cases Committee RCAF "R" Depot

The following figures have been selected from a postwar summary entitled "Statistics – Special Cases Committee" of the RCAF "R" Depot, which covers the period September 1944 to September 1945, when the "R" Depot disposed of 627 cases (tables B.1 and B.2).[1] In addition to those cases, the Reselection Board dealt with 448 cases from 11 September 1944 to 30 September 1945.[2]

Table B.1
Classification of Cases according to the Air Ministry [LMF] Memorandum and RAF Policies, When They Reached the RCAF "R" Depot

Classification	Number	Per cent
5(i) [LMF]	59	9.4
5(ii) [Medical "WS"]	35	5.6
Misconduct	54	8.6
Inefficiency	471[a]	75.1
Unclassified	8	1.3
Totals	627	100.0

[a] "This total includes 41 cases of airmen where the inefficiency, though falling short of wavering, was of such a character that reduction in rank immediately prior to discharge was recomended and approved. In corresponding cases of culpable inefficiency on the part of officers, reduction in rank was not practicable."

Table B.2
Cases Categorized by the Special Cases Committee under "AFCO 1/45" after
Their Arrival at the "R" Depot

Paragraph and Classification under "AFCO 1/45"	Number	Per cent
1(a) Wilful Attempt to Escape Aircrew Hazards	40	6.4
2(a) Exonerated – exceptional stress	5	0.8
3(b) Nervous Ailment – no resolute effort	5	0.8
3(c) Exonerated – nervous ailment – resolute effort	32	5.1
4(c) Inefficiency	457	72.9
5(b) Misconduct – breach of flying discipline	4	0.6
5(c) Other Misconduct	51	8.1
Unclassified	33	5.3
Totals	627	100.0

APPENDIX C
RCAF Aircrew NP Ratios in Bomber Command, 1944–45

This appendix outlines how NP ratios were calculated by the author for all of Bomber Command and for RCAF aircrew in Bomber Command for 1944–45. As with the other statistics relating to psychological casualties presented in this book, these should be treated as approximations for the purpose of determining orders of magnitude.

From 10 February 1944 to 9 February 1945 RAF neuropsychiatrists labelled 306 aircrew LMF and diagnosed 2,910 as suffering from a "neurosis" (for a total of 3,216 cases). Bomber Command accounted for 30.3 per cent of all RAF cases of "recognized neurosis" during the period under discussion, and thus about 975 of the 3,216 LMF and "neurosis" cases can be allocated to it.[1] Bomber Command operational and non-operational aircrew casualties from 3 September 1944 to 8 May 1945 amounted to 11,444 (adjusted to an annual figure of 17,080).[2] Combining these figures results in an NP ratio (the number of LMF and "neurosis" cases divided by the adjusted number of casualties) of 5.7 per cent.

From September 1944 to September 1945, 94 aircrew classified as waverers reached the RCAF "R" Depot.[3] If the RCAF had the same ratio of LMF to "neurosis" cases as that found by RAF neuropsychiatrists for the general aircrew population, then there would have been 895 cases of "neurosis" among the RCAF aircrew at Warrington, and LMF and "neurosis" cases combined would have totalled 989. I was unable to locate any RCAF statistics for LMF or neurosis by Command, but using the RAF figure of 30.3 per cent for Bomber Command results in approximately 300 LMF and "neurosis" cases that can be allocated to RCAF aircrew from Bomber Command. RCAF Bomber Command operational and non-operational aircrew casualties from 3

September 1944 to 8 May 1945 totalled 2,904 (adjusted to an annual figure of 4,334),[4] and the RCAF NP ratio was thus 6.9 per cent.

Notes

All archival references are to the National Archives of Canada unless otherwise specified.

CHAPTER ONE

1 Contemporary Civilization Staff, Columbia College, *Introduction*, 10–12 (quotation at 11).
2 Gould, *The Mismeasure of Man*, 21.
3 B.D. Hunt and R.G. Haycock, "Introduction" and "Section 1: 1904–1945," in Hunt and Haycock, eds., *Canada's Defence*, 1, 6 (quotation at 1).
4 Desmond Morton, "'Junior but Sovereign Allies': The Transformation of the Canadian Expeditionary Force," in Hunt and Haycock, eds. *Canada's Defence*, 32.
5 Adrian W. Preston, "Canada and the Higher Direction of the Second World War 1939–1945," in Hunt and Haycock, eds., *Canada's Defence*, 99.
6 Morton, *Military History of Canada*, 225; Robert Bothwell, "'Who's Paying for Anything These Days?' War Production in Canada 1939–45," in Dreisziger, ed., *Mobilization*, 68.
7 Bothwell, "'Who's Paying for Anything These Days?'" in Dreisziger, ed., *Mobilization*, 59.
8 The phrase "aerodrome of democracy" was apparently coined by Lester B. Pearson in 1942: Hatch, *Aerodrome of Democracy*, iv.
9 Quoted in Roberts, *There Shall Be Wings*, 132–3.

10 Terraine, *The Right of the Line*, 682.

11 Stacey, *Arms, Men and Governments*, 66, 259, 305; Feasby, ed., *Official History*, 512.

12 "Manpower" was the term used in World War II. The contemporary usage has been retained in this book, even though human-resource questions involved both men and women. There were 45,423 women in the Canadian armed forces in World War II, 17,018 of whom served in the Royal Canadian Air Force (RCAF) Women's Division. All RCAF aircrew were male: Stacey, *Arms, Men and Governments*, 416, 590.

13 Marc Milner, "Introduction," in Milner, ed., *Canadian Military History*, 2.

14 Ruggle, review of *Battle Exhaustion*, 46.

15 Van Creveld, *Technology and War*, 229. Emphasis in original.

16 Van Creveld, *Technology and War*, 232.

17 Van Creveld, *Command in War*, 275.

18 McNeill, *The Pursuit of Power*, 317, 339, 355–6.

19 Van Creveld, *Technology and War*, 224.

20 Copp, "From Flanders Fields to Sarajevo."

21 Schwartz, "The Creation and Destruction of Value," 7, 12.

22 Webster and Frankland, *Strategic Air Offensive against Germany*, 4: 446.

23 Stephen Harris and Norman Hilmer, "The Development of the Royal Air Force, 1909–1945," in Jordan, ed., *British Military History*, 345.

24 Robinson, *The Dangerous Sky*, 100.

25 The 1911 census showed Canada's population as 7,206,643; by 1921 it had risen to 8,788,483. My calculation of the percentage of the national population in uniform is based on an assumed total Canadian population of 7,997,563 (the midpoint of the 1911 and 1921 census figures).

26 Morton, *Canada and War*, 82, 84.

27 Stacey, *Arms, Men and Governments*, 66, 590; Morton, *Military History of Canada*, 185.

28 Burns, *Manpower*, 142.

29 Van Creveld, *Supplying War*, 236.

30 Wise, *Canadian Airmen*, 168, 283–4.

31 Wise, *Canadian Airmen*, 600.

32 Douglas, *Creation of a National Air Force*, 61–2. For reasons of convenience, the designations Royal Flying Corps (RFC)/Royal Air Force (RAF) and Canadian Air Force (CAF)/RCAF will be used in this book where required. Unless otherwise indicated, a reference to the RFC/RAF is presumed to include the Royal Naval Air Service (RNAS).

33 Precise figures are difficult to ascertain. Sir A.E. Kemp, appointed minister of militia and defence in 1916, "estimated that 25 per cent of all RAF flying personnel ... were Canadian": Wise, *Canadian Airmen*, 593, 597 (quotation at 597).

34 Douglas, *Creation of a National Air Force*, 31, 91.

35 Douglas, *Creation of a National Air Force*, 134, 138 (quotation at 134).

36 Douglas, *Creation of a National Air Force*, 62–3, 120, 145–6.

37 Douglas, *Creation of a National Air Force*, 372.

38 Leonard Birchall quoted in Foster, *For Love and Glory*, 7. The Actual strength of the RCAF on 5 September 1939 (all ranks) was 4,153, but this included recently mobilized auxiliary personnel: Douglas, *Creation of a National Air Force*, 343.

39 Douglas, *Creation of a National Air Force*, 134.

40 Roberts, *There Shall Be Wings*, 116; Douglas, *Creation of a National Air Force*, 343.

41 Quoted in Bothwell, "'Who's Paying for Anything these Days?'" in Dreisziger, ed., *Mobilization*, 59.

42 Morton, *Military History of Canada*, 178.

43 Plumptre, *Mobilizing Canada's Resources for War*, xiv.

44 Bothwell, "'Who's Paying for Anything these Days?'" in Dreisziger, ed., *Mobilization*, 59. According to Bothwell, Skelton repeatedly warned of the dangers of committing Canada to the war against the Axis.

45 Morton, *Military History of Canada*, 237–8.

46 Ward, ed., *A Party Politician*, 201.

47 Norman Rogers quoted in Douglas, *Creation of a National Air Force*, 344.

48 Air Vice-Marshal G.M. Croil, Chief of the Air Staff (CAS), quoted in Douglas, *Creation of a National Air Force*, 344.

49 Robinson, *The Dangerous Sky*, 77–80; Fredette, *The Sky on Fire*, 215.

50 Barris, *Behind the Glory*, 34.

51 Terraine, *The Right of the Line*, 258; Douglas, *Creation of a National Air Force*, 247.

52 Douglas, *Creation of a National Air Force*, 221.

53 Roberts, *There Shall Be Wings*, 134.

54 Douglas, *Creation of a National Air Force*, 194.

55 Douglas, *Creation of a National Air Force*, 193. The senior RAF officer was Air Vice-Marshal Charles Portal.

56 Ward, ed., *A Party Politician*, 199, 235 (quoting Sir John Slessor at 235).

57 Morton, *Military History of Canada*, 202.

58 Quoted in Terraine, *The Right of the Line*, 259.

59 Terraine, *The Right of the Line*, 290, 504.

60 Longmate, *The Bombers*, 175.

61 Terraine, *The Right of the Line*, 290–1, 471, 504.

62 Longmate, *The Bombers*, 217.

63 Webster and Frankland, *Strategic Air Offensive against Germany*, 3: 79–80.

64 Kingston-McCloughry, "Leadership," 208. See also Hastings, *Bomber Command*, 278, 282–3, 291–2.

65 Terraine, *The Right of the Line*, 765 n. 18.

66 Terraine, *The Right of the Line*, 682.

67 Webster and Frankland, *Strategic Air Offensive against Germany*, 4: 440. This number does not include Canadians serving with the RAF.

68 Terraine, *The Right of the Line*, 536 (quoting Sir Arthur Harris).

69 Sir John Slessor quoted in Terraine, *The Right of the Line*, 260.

70 Terraine, *The Right of the Line*, 522; Air Vice-Marshall D.C.T. Bennett, "Foreword," in Thompson, *Lancaster to Berlin*, n.p. Bennett was the leader of the Pathfinders, an élite Bomber Command group that guided the attacking aircraft to the target, and marked it for them.

71 Ward, ed., *A Party Politician*, 202, 235–6.

72 Stacey, *A Date with History*, 257.

73 Carter, *RAF Bomber Command and No. 6 [Canadian] Group*, 132, 162.

CHAPTER TWO

1 According to a letter of 30 January 1919 from a Mr. W. Smith of Belleville, Ontario, to the Minister of Militia and Defence, this slogan was used in the "appeal for airmen in the Fall of 1916." Smith claimed that the appeal induced him to sell his business and join the Royal Canadian Naval Air Service at the age of forty-four: RG 24, vol. 2043, HQ 6978-2-131, vol. 10, ff. 50–1.

2 Deuteronomy 20: 5–9.

3 Deuteronomy 20: 8. Emphasis in original.

4 Edgar M. Johnson, "Foreword," in Gal and Mangelsdorff, eds., *Handbook*, xxi; A. David Mangelsdorff and Reuven Gal, "Overview of Military Psychology," in Gal and Mangelsdorff, eds., *Handbook*, xxvi.

5 Gould, *The Mismeasure of Man*, 24, 157, 174, 188, 194, 196–7.

6 The strongest argument for the perfectibility of aircrew selection methods is presented in D.R. Hunter, "Aviator Selection," in Wiskoff and Rampton, eds., *Military Personnel Measurement*, 158–61. A counter-argument may be found in English and Rodgers, "Déjà Vu?" 35–6, 39–43, 45.

7 Alan Jones, "The Contribution of Psychologists to Military Officer Selection," in Gal and Mangelsdorff, eds., *Handbook*, 66.

8 Directorate of History, Department of National Defence (DHIST), "RCAF Personnel History," 2: 415; Douglas, *Creation of a National Air Force*, 293.

9 Wise, *Canadian Airmen*, 644.

10 Macpherson, *History*, 2: 115; Rexford-Welch, ed., *Royal Air Force Medical Services*, 1: 3–5.

11 Birley, "Principles ... Lecture I," 1147–8; Jones, *Flying Vistas*, 40.

12 Macphail, *The Medical Services*, 20–4, 158, 161, 193, 320; Macpherson, *History*, 1: 118, 120, 122–5, 129; Mitchell and Smith, *History*, 12.

13 Haycock, *Hughes*, 179, 200; Macphail, *The Medical Services*, 160–1.

14 Bruce, *Politics*, 40, 289–303, 321; Macphail, *The Medical Services*, 156–69.

15 Both the RFC and the RNAS started limited recruiting efforts in Canada in 1915. Before then, only British-born candidates were accepted into the two air services: Wise, *Canadian Airmen*, 29–32, 36–9, 76.

16 Sullivan, *Aviation in Canada*, 147, 151; Wise, *Canadian Airmen*, 87.

17 Sullivan, *Aviation in Canada*, 143, 149. Brigadier-General C.G. Hoare, General Officer Commanding (GOC) of RFC Canada, requested and was granted permission in March 1917 for standing medical boards to examine Cadet (Pilots) according to RFC medical standards: RG 24, vol. 2040, HQ 6978-2-131, vol. 1, ff. 149, 156, Hoare to Secretary Militia Council (SMC), 8 March 1917; SMC to Hoare, 16 March 1917. In 1918 there was a lack of doctors outside Toronto qualified to administer the "Barany-Chair Tests" of equilibrium. As a result, all candidates had to travel to Toronto for final testing: RG 24, vol. 2042, HQ 6978-2-131, vol. 6, f. 171, Hoare to Gwatkin, 13 May 1918.

18 Sullivan, *Aviation in Canada*, 135; Wise, *Canadian Airmen*, 84; RG 24, vol. 2040, HQ 6978-2-131, vol. 1, ff. 66, 68, 69, Adjutant-General to GOC Militia District (MD) No. 2, 5 February 1917; Assistant Director of Medical Services (ADMS) MD No. 2 to Director General of Medical Services (DGMS), 6 February 1917; DGMS to ADMS MD No. 2, 9 November 1917.

19 While serving on exchange with the RAF from 1978 to 1981, I had to take the whisper test as part of my annual medical, for my RAF station had not yet acquired a hearing-test machine (standard in the Canadian Forces at the time). In one instance, a young Scottish nurse, with a very thick Highland accent, stood at one end of the corridor, and I at the other. She whispered something that sounded like "coo" to me, and I dutifully repeated it. After several repetitions of the same word, at increasing volume, until we were almost shouting, we agreed that she had said "cow." One or two equally unsuccessful attempts with other words followed, until the test was brought to a conclusion on the understanding that my hearing was normal but our languages different.

20 RG 24, vol. 2041, HQ 6978-2-131, vol. 3, ff. 302–3, Hoare to Gwatkin, 20 December 1917; f. 355, "Medical Requirements for Cadets," 28 November 1917; vol. 2042, HQ 6978-2-131, vol. 8, f. 81, "Medical Requirements for Cadets" (summer 1918).

21 RG 24, vol. 2041, HQ 6978-2-131, vol. 5, ff. 193–208, "Strength RFC Canada," 14 February 1918.

22 RG 24, vol. 2041, HQ 6978-2-131, vol. 3, ff. 315–16, "Officers with the RFC in Canada," 16 November 1917; vol. 4, f. 5, "Officers with the RFC in Canada," 16 December 1917; vol. 5, ff. 174–92, "Strength of RFC

Canada," 28 January 1918; ff. 193–208, "Strength of RFC Canada," 28 January 1918; f. 216, "Strength of RFC Canada," 7 March 1918; vol. 2042, HQ 6978-2-131, vol. 6, ff. 204, 216, "Officers on Strength with the RAF Canada," 7 May 1918; vol. 8, ff. 50–62, 64–75, 139–70, "Officers on Strength with the RAF Canada," 7 June, 7 July, 13 September 1918.

23 RG 24, vol. 2041, HQ 6978-2-131, vol. 4, ff. 146–56, Correspondence relating to the secondment of Canadian Army Medical Corps (CAMC) doctors to RFC Canada, 29 December 1917–20 February 1918.

24 Sullivan, *Aviation in Canada*, 108–12.

25 Hamel and Turner, *Flying*, 324–8.

26 Sullivan, *Aviation in Canada*, 112–13.

27 Sullivan, *Aviation in Canada*, 118.

28 RG 24, vol. 2042, HQ 6978-2-131, vol. 6, f. 171, Hoare to Gwatkin, 13 May 1918; Birley, "Principles ... Lecture III," 1254; Armstrong, *Principles*, 50–1; Robinson, *The Dangerous Sky*, 86–8.

29 RG 24, vol. 2041, HQ 6978-2-131, vol. 3, f. 304 Officer-in-Charge No. 2 District RFC to Assistant Adjutant-General MD No. 12, 14 November 1917; Wise, *Canadian Airmen*, 87.

30 Haycock, *Hughes*, 200, 202.

31 Sullivan, *Aviation in Canada*, 149.

32 Sullivan, *Aviation in Canada*, 149. Partial reimbursement, totalling £75, was offered to successful applicants by the British services, and some candidates received financial support from Canadian sources: Wise, *Canadian Airmen*, 31, 38–9.

33 Wise, *Canadian Airmen*, 87.

34 Sullivan, *Aviation in Canada*, 149.

35 Chajkowsky, *History of Camp Borden*, 71.

36 Wise, *Canadian Airmen*, 87–8.

37 Mauer, *Aviation in the U.S. Army*, xxiv.

38 RG 24, vol. 2041, HQ 6978-2-131, vol. 3, f. 296, Colonel, Adjutant-General Organization to Adjutant-General Canadian Militia, 15 November 1917.

39 Wise, *Canadian Airmen*, 31.

40 Some exceptions seem to have been made, however. Norman Macmillan claimed that he knew of four pilots in the RFC who were natives of India; he recalled that three of them were killed during the war: Macmillan, *Offensive Patrol*, 70. A.J. Insall recounted that he flew with a Maori RFC pilot in 1915: Insall, *Observer*, 58.

41 RG 24, vol. 2042, HQ 6978-2-131, vol. 8, f. 114, untitled newspaper clipping, n.d.

42 RG 24, vol. 2041, HQ 6978-2-131, vol. 5, f. 149, Duke to Minister of Militia and Defence, 10 January 1918.

43 H.C.G. Matthews, "The Liberal Age," in Morgan, ed., *Oxford Illustrated History of Britain*, 511.

44 G[rey], "A Question of Ethnology," 227–8; Raleigh, *War in the Air*, 1: 204–5. Walter Raleigh wrote the first of the six volumes of *War in the Air*; H.A. Jones wrote the others.

45 Ruck, "Blacks," 11. In October 1940 the RCAF Air Council stated, in reply to a request to commission a "British Indian" applicant, that "Regulations prescribe that candidates for a commission must be of pure European descent": RG 24, vol. 5172, HQC 15-1-136, Copy of "Summary-Air Members' Meeting," 18 October 1940.

46 A good summary of this debate may be found in Robinson, *The Dangerous Sky*, 82–9.

47 Greenhous, *A Rattle of Pebbles*, 55; McCudden, *Flying Fury*, 155.

48 Winter, *The First of the Few*, 24–5.

49 For example, the RAF's Master-General of Personnel, Brigadier-General William Sefton Brancker, commented in September 1918 that Canadian pilots "'are a better type than we are getting in England now'": Wise, *Canadian Airmen*, 112–13.

50 Wise, *Canadian Airmen*, 327. Harold Price, a Canadian commissioned as a Second Lieutenant in the RFC, said on his arrival in England in December 1916 that "in a country like this one needs five meals a day to keep warm. It is foggy and cruelly bitter": Greenhous, *A Rattle of Pebbles*, 164.

51 Winter, *The First of the Few*, 22–3.

52 Wise, *Canadian Airmen*, 91, 95.

53 Bingham, *An Explorer in the Air Service*, 11–22.

54 Henmon, "Air Service Tests," 103–4. This point was brought to Bingham's attention during his May 1917 visit to RFC Canada, where he was told that the pilot was not "'a flying chauffeur,'" but a "'modern cavalry officer'" or a "'knight of old'": Bingham, *An Explorer in the Air Service*, 16–17.

55 Jones, *Flying Vistas*, 182–3, 185–6, 188.

56 Armstrong, *Principles*, 30–1, 41.

57 Rexford-Welch, ed., *Royal Air Force Medical Services*, 1: 3–5; Macpherson, *History*, 2: 115.

58 "The Medical Service of the Air Force," 419; Rexford-Welch, ed., *Royal Air Force Medical Services*, 1: 6.

59 RG 24, vol. 2041, HQ 6978-2-131, vol. 5, f. 216, "Strength of RFC Canada," 7 March 1918.

60 RG 24, vol. 2043, HQ 6978-2-131, vol. 9, ff. 177–8, Parker to O'Reilly, 24 December 1918.

61 RG 24, vol. 2043, HQ 6978-2-131, vol. 9, f. 179, O'Reilly to Fotheringham, 15 January 1919.

62 RG 24, vol. 2043, HQ 6978-2-131, vol. 9, ff. 166–7, Fotheringham to Chief of the General Staff (CGS), 8 January 1919 and minutes; Mac-Phail, *The Medical Services*, 319, 404.

63 Hitchins, *Air Board*, 38, 94.

64 Douglas, *Creation of a National Air Force*, 92.

65 About twenty student pilots a year graduated from 1932 to 1940: see the figures in Hitchins, *Air Board*, 184, 249, 266, 295–6, 306, 307, 321, 335–6, 363, 386, 387–8.

66 DHIST, "RCAF Personnel History," 1: 15. On 14 October 1939 the authorized strength of the RCAF was 16,500 officers and men. This was raised on 23 February 1940 to 44,500, and on 15 April 1941 to 92,000.

67 The RCAF always had an excess of recruits, which at times was "embarrassingly large": Hatch, *Aerodrome of Democracy*, 181–7. See also DHIST, Directorate of Manning, RCAF, "Manning the RCAF," 4. The other major bottlenecks were a lack of training facilities and a lack of aircraft.

68 DHIST, "RCAF Personnel History," 3: 654–5.

69 RG 24, vol. 3303, HQ 1260-1-26, "Air Force Manning Orders M. 24/2," Appendix A, 31 December 1941.

70 RG 24, vol. 4947, HQ 895-8/12, vol. 1, f. 30, "Schedule by Day," n.d. (January [?] 1940); f. 67, Hunter to Air Member for Personnel (AMP), n.d. (February [?] 1940); f. 99, Morgan Smith to Staff Officer Medical Services (SOMS) (Air), 18 July 1940; vol. 2, f. 15, Tice to Air Officer Commanding (AOC) No. 4 Training Command, 16 December 1940.

71 Feasby, ed., *Official History*, 2: 57.

72 RG 24, vol. 4947, HQ 895-8/12, vol. 1, f. 109, Hunter to AMP, 24 July 1940.

73 Douglas, *Creation of a National Air Force*, 624.

74 DHIST, "History of the AMP Division," 12; DHIST, "RCAF Personnel History," 1: 12; RG 24, vol. 5172, HQC 15-1-136, "Minutes of AMP Directors Meeting," 9 December 1941.

75 Ettinger, *History*, 9; Swettenham, *McNaughton*, 1: 331; National Research Council of Canada (NRC), *History*, 1: 4–5 (quotation at 4). The NRC's other matters of pressing concern were "respiration, gas metabolism and rare gases at lower pressures" and devising improved "emergency oxygen apparatus": NRC, *History*, 4–5.

76 MG 28, I 161, vol. 17, file 17-9, Bott to McEachran, 21 November 1939; Wright, "CPA [Canadian Psychological Association]," 115. Bott had lobbied for the post, and established a personal relationship with McNaughton and McNaughton's successor.

77 "News and Comment" (October 1940), 6; NRC, *History*, 2–3, 26–7, 60–70.

78 Myers, "Bott," 292–302.

79 MG 28, Chant transcript, 32; Ferguson interview with the author.

80 MG 28, Chant transcript, 33.

81 Wright, "CPA," 116.

82 Bott and Cosgrave, "Preliminary Studies," 79.

83 M.J. Wright and C.R. Myers, "Introduction," in Wright and Myers, eds., *History*, 17.

84 MG 28, I 391, file 17-9, "Minutes of a Regional Meeting of CPA," 11 September 1939.

85 NRC, *History*, 60.

86 Bott and Cosgrave, "Preliminary Studies," 65, 68–9, 74–5.

87 Bott, "Report," A-1 to A-3; NRC, *History*, 60.

88 MG 28, I 161, file 17-9, Humphrey to McEachran, 21 November 1939; Bott to McEachran, 2 December 1940.

89 Bott and Cosgrave, "Preliminary Studies," 65–9, 77–9. Bott used the World War I aviator's slang "guts" to describe the aviator's "tendency to preserve his drive under pressure" and his ability to endure stress: RG 24, vol. 4947, HQ 895-8/12, Part 1, ff. 168–9, "Syllabus of First Course."

90 Bott, "Report," A-1 to A-3; DHIST, "RCAF Personnel History," 3: 655–6, 669–79.

91 For some of Bott's reports see Bott, "Memorandum"; Bott and Cosgrave, "Preliminary Studies"; MG 28, I 161, vol. 17, file 17-9, Bott to McEachran, 21 November 1939 and 2 December 1940.

92 RG 24, vol. 5383, HQS 47-6-3, Associate Committee on Aviation Medical Research, "Proceedings of the Executive," 29 July 1941, Appendix B. For a more complete account see English, "Canadian Psychologists."

93 Vernon and Parry, *Personnel Selection*, 71–2; MG 28, Myers transcript, 87–90.

94 RG 24, vol. 3204, HQ 145-5-7, AMP to Air Member for Organization (AMO), 23 January, 13 May, 23 June 1942.

95 Williams, "Chant," 86; MG 28, Chant transcript, 1–24; C.R. Myers, "Psychology at Toronto," in Wright and Myers, eds., *History*, 86; MG 28, I 161, vol. 20, "Publications of S.N.F. Chant."

96 RG 24, vol. 3204, HQ 145-5-4, AMP to AMO, 13 May, 23 June 1942; copy of CAS C.19-15-2, 22 November 1943.

97 MG 28, Belyea transcript, 56.

98 Williams, "Chant," 86.

99 "News and Comment" (February 1946), 15.

100 DHIST, "BCATP [British Commonwealth Air Training Plan] Flying Training: The Visual Link Trainer," 1.

101 Hitchins, *Air Board*, 369.

102 DHIST, "BCATP Flying Training: The Visual Link Trainer," 2, 5, 9, 12, 16, 23.

103 RG 24, box 2194, HQ 473-1, vol. 3, "The Link Test of Flying Aptitude, Standard Sequence of Instruction, Authorized by the Directorate of Personnel Selection and Research, December, 1943."

104 DHIST, "BCATP Flying Training: The Visual Link Trainer," 19–21.

105 Bott, "Memorandum," 10–11.

106 DHIST, "BCATP Flying Training: The Visual Link Trainer," 5–9; MG 28, Williams transcript, 60–4; RG 24, vol. 2193, HQ 927-11-35, "Outline of the Underlying Principles in Using the Visual Link Trainer for Pilot Selection," n.d. The "Outline" was sent to Air Force Headquarters (AFHQ) in September 1941, and it appears to be the report that aroused Chant's interest in the Visual Link Trainer.

107 Signori, "The Arnprior Experiment"; DHIST, "BCATP Flying Training: The Visual Link Trainer," 34. Myers recalled that he was the person principally responsible for the RAF's adoption of flight grading: MG 28, Myers transcript, 89–93.

108 Staffs, "Research Program," 307–21; DHIST, "BCATP Flying Training: The Visual Link Trainer," 30, 34–5.

109 DHIST, "BCATP Flying Training: The Visual Link Trainer," 19–22.

110 Leckie to AOCS Nos. 1, 2, 3, and 4 Training Commands, 2 April 1943, in DHIST, "BCATP Flying Training: The Visual Link Trainer," 18–19; Douglas, *Creation of a National Air Force*, 242.

111 The same technical problems that beset the Link Trainer during the war continued into the 1950s and 1960s. Because of a limited number of spaces in specialist courses, not all technicians assigned to Link Trainer maintenance had received the training necessary to qualify them to repair the apparatus. And since a posting to Link Trainer maintenance was seen as a break in an instrument technician's usual career path, some were not interested in the assignment. In addition, the Link Trainer hydraulic system was particularly susceptible to changes in humidity, and as few of the rooms where the machines were kept were air-conditioned, performance and test results could vary with climatic conditions: Gushue interview with the author. Fluctuations in the voltage of the electricity supplied to Link Trainer installations could also affect performance: RG 24, box 2194, HQ 473-1, vol. 3, Chant to Commanding Officer (CO) No. 6 Initial Training School (ITS), 9 December 1943.

112 DHIST, "BCATP Flying Training: The Visual Link Trainer," 28.

113 D.M.S. (Air), "Factors in Aircrew Selection," 2. The correlation coefficient is a measure of the relationship between two variables, in this case Link Trainer scores and success in pilot training. "In psychological research, a correlation coefficient of .60 or more is judged to be quite high. Correlations in the range from .20 to .60 are of practical and theoretical value and useful in making predictions": Atkinson et al., *Introduction to Psychology*, 23–4.

114 D.R. Hunter, "Aviator Selection," in Wiskoff and Rampton, eds., *Military Personnel Measurement*, 152–3. In Hunter the terms "validity," "validity coefficients," and "coefficient of correlation" are used synonymously (132).

115 Signori, "The Arnprior Experiment"; MG 28, Signori transcript; Hunter, "Aviator Selection," 144–5.

116 DHIST, "BCATP Flying Training: The Visual Link Trainer," 29–32, 34–5.

117 Thomas F. Hilton and Daniel L. Dolgin, "Pilot Selection in the Military of the Free World," in Gal and Mangelsdorff, eds., *Handbook*, 88.

118 Chant, "Psychology," 37; MG 28, Myers transcript, 89.

119 Gould, *The Mismeasure of Man*, 24, 27.

120 Gould, *The Mismeasure of Man*, 230–2; Hunter, "Aviator Selection," in Wiskoff and Rampton, eds., *Military Personnel Management*, 132, 134. The correlation coefficient on the RCAF test was .06. Postwar research confirmed this result.

121 Chant, "Psychology," 36–7.

122 Feasby, ed., *Official History*, 2: 457–9.

123 McIntosh, *Terror in the Starboard Seat*, 10–11.

124 Feasby, ed., *Official History*, 2: 458, 460.

125 Psychologists were not the only specialists to vary standards according to demand. Before the war, RAF doctors modified the standard for colour vision when it seemed that stringent criteria were eliminating too many aircrew candidates: Livingston, *Fringe of the Clouds*, 184.

126 MG 28, Williams transcript, 65–6.

127 See, for example, Thorndike, "Selection of Military Aviators – III," 29–31.

128 Staff Psychological Section, "Psychological Activities," 43-50.

129 Jones, *Flying Vistas*, 192.

130 Thorndike, "Selection of Military Aviators – Mental and Moral Qualities," 14.

CHAPTER THREE

1 Air Ministry, *Flying Training*, 1; Lewis, *Aircrew*, 100.

2 Air Ministry, *Flying Training*, 1, 84, 169, 209.

3 Terraine, *The Right of the Line*, 504.

4 Air Ministry, *Flying Training*, 199.

5 Public Record Office (United Kingdom) (PRO), AIR 2/4935, "The Operational Tour-Minutes of Meeting Held on Thursday, 4th February, 1943.

6 Air Ministry, *Flying Training*, 175.

7 See, for example, Douglas, *Creation of a National Air Force*, 191–293; Air Ministry, *Flying Training*, 108, 166; Terraine, *The Right of the Line*, 81–91.

8 Hawkers, *Hawker, V.C.*, 27–8.

9 Strange, *Recollections*, 18–19.

10 Until 1 January 1914 the test specified a glide from 100 metres: "The New Certificate Tests," 227.

11 Insall, *Observer*, 30; Lewis, *Wings over the Somme*, 17; Wise, *Canadian Airmen*, 29.

12 Wise, *Canadian Airmen*, 84.

13 Jones, *War in the Air*, 5: 424; Strange, *Recollections*, 131–2.

14 Raleigh, *War in the Air*, 1: 441–2; Jones, *War in the Air*, 3: 292–7.

15 Jones, *War in the Air*, 5: 426. Henmon, "Air Service Tests," 103–4, gives attrition rates of 15 per cent in ground-school training and 6 per cent in flying training. See Wise, *Canadian Airmen*, 106–7 and 645–9 on the difficulties historians face in collecting meaningful RFC personnel statistics from World War I.

16 Taylor, *Sopwith Scout 7309*, 23; Vee, *Flying Minnows*, 32, 57–8. Voss published his book under the pseudonym Roger Vee.

17 Wise, *Canadian Airmen*, 106.

18 Vee, *Flying Minnows*, 44–5. Emphasis in original.

19 Wise, *Canadian Airmen*, 106.

20 Lewis, *Sagittarius Rising*, 256.

21 Bingham, *An Explorer in the Air Service*, 162, 174–5.

22 Jones, *War in the Air*, 5: 429–30.

23 Macmillan, *Into the Blue*, 27, 141; Strange, *Recollections*, 16–18; Jones, *War in the Air*, 5: 430.

24 Jones, *War in the Air*, 5: 429–32, 434, 448.

25 Lewis, *Wings over the Somme*, 115; Macmillan, *Into the Blue*, 183.

26 Stratton et al., "Psychological Tests," 420–2; Bingham, *An Explorer in the Air Service*, 147.

27 Sullivan, *Aviation in Canada*, 220; Wise, *Canadian Airmen*, 105.

28 The major differences between the Canadian and British systems were due to the fact that RFC Canada used the JN4 aircraft for training instead of the Avro 504. Unlike the Avro 504, the JN4 did not have a rotary engine. This made its handling characteristics very different, and necessitated modifications to the Gosport syllabus: Sullivan, *Aviation in Canada*, 220; Wise, *Canadian Airmen*, 105.

29 Sullivan, *Aviation in Canada*, 211.

30 Sullivan, *Aviation in Canada*, 220–3; Wise, *Canadian Airmen*, 105.

31 Sullivan, *Aviation in Canada*, 104. The same concept of a separate standards cell or "examining party," as it was called, was also applied in ground-school training at RFC Canada's School of Military Aeronautics: 165.

32 Sullivan, *Aviation in Canada*, 217–18.

33 Sullivan, *Aviation in Canada*, 135–6.

34 Sullivan, *Aviation in Canada*, 135–6, 155–9, 162; Wise, *Canadian Airmen*, 84–7.

35 Sullivan, *Aviation in Canada*, 76–81, 162–5, 170–9, 180–9; Wise, *Canadian Airmen*, 85–6.

36 Wise, *Canadian Airmen*, 99.

37 See the remarks on Hugh Trenchard in chapter 6.

38 Wise, *Canadian Airmen*, 107–8.

39 Even as late as 1918 engines were quite undependable. Tests conducted by the RAF's Central Flying School (CFS) on an SE5a showed that there would be engine trouble on one in every six flights: Winter, *The First of the Few*, 111.

40 Wise, *Canadian Airmen*, 100–1, 118.

41 Wise, *Canadian Airmen*, 113, 117–18.

42 Douglas, *Creation of a National Air Force*, 52; Hitchins, *Air Board*, 85.

43 Hitchins, *Air Board*, 85–6, 249–50, 266, 375.

44 Based on figures from Hitchins, *Air Board*, 266, 295–6, 306–7, 321, 335–6, 363, 386, 387–8.

45 Hitchins, *Air Board*, 184, 249. The attrition rates for the non-commissioned officer (NCO) courses that began in 1936 cannot be computed from the data in Hitchins, as no intake numbers are given. During the years 1923–36 the Provisional Pilot Officer (PPO) and NCO schemes produced a total of about 170 pilots: 100, 123, 146, 165, 184, 208, 225, 236, 250, 266, 281, 296, 306, 321, 336. It is not possible to determine the number of pilots produced by the PPO scheme after 1936 because the figures in Hitchins are distorted by other pre-war pilot-training expansion schemes.

46 Douglas, *Creation of a National Air Force*, 62, 93; Hitchins, *Air Board*, 184.

47 Douglas, *Creation of a National Air Force*, 75; Hitchins, *Air Board*, 400.

48 From 1928 to 1932, ninety-five civilian instructors attended the RCAF courses. There are no attendance figures for the period after 1932: Hitchins, *Air Board*, 208, 224, 236, 249, 266.

49 Hitchins, *Air Board*, 280 n.6. In 1934 seven instructors took RCAF certification tests: 296. The testing and certification program continued until the outbreak of war: 365.

50 Hitchins, *Air Board*, 291. For the period beginning in 1935 Hitchins uses the term "preliminary training": 306, 335.

51 Hitchins, *Air Board*, 364, 376, 386.

52 Like the RAF, the RCAF still conducted much of its intermediate and advanced training at the squadron level: Air Ministry, *Flying Training*, 2; Douglas, *Creation of a National Air Force*, 148–9; Hitchins, *Air Board*, 306, 321–2, 335, 365.

53 Hatch, *Aerodrome of Democracy*, 116.

54 PRO, Air Historical Board (AHB), "Flying Training," 59.

55 PRO, AHB, "Flying Training," 69.

56 PRO, AHB, "Flying Training," 60–1.

57 PRO, AHB, "Flying Training," 72. Tedder had been intimately involved with training during the interwar years, first as commander of RAF Reserves, and in 1934–36 as Director of Training at the Air Ministry: 29; Conrad, *Training for Victory*, 2; Terraine, *The Right of the Line*, 43.

58 PRO, AHB, "Flying Training," 64–5.

59 PRO, AHB, "Flying Training," 64–5, 69.

60 Douglas, *Creation of a National Air Force*, 371.

61 Hatch, *Aerodrome of Democracy*, 108–9; Douglas, *Creation of a National Air Force*, 240, 247, 291; Air Ministry, *Flying Training*, 84, 121–2, 150, 155, 213–14, 225.

62 Air Ministry, *Flying Training*, 167–8. Some instructors posted to Operational Training Units (OTUS) had not even qualified as aircraft captains on their operational tours.

63 Air Ministry, *Flying Training*, 152–3, 166–7, 210.

64 Air Ministry, *Flying Training*, 178–80, 195. It was reported in *Flying Training* that a student pilot during the one-pilot regime would receive eighty hours of OTU flying, against thirty hours under the former regime: 195.

65 Detailed descriptions of this process can be found in Douglas, *Creation of a National Air Force*, 220–8, and in Hatch, *Aerodrome of Democracy*, 33–45.

66 Douglas, *Creation of a National Air Force*, 228–9, 626. The RCAF appears to have been ahead of the RAF in this area, as the RAF did not centralize its flying training until it formed a Training Command and created the position of Air Member for Training (AMT) in mid-1940: Air Ministry, *Flying Training*, 68–9, 95–6. The RCAF had formed an Air Training Command in 1938. Leckie began to coordinate training in early 1940: Douglas, *Creation of a National Air Force*, 228–9.

67 Hatch, *Aerodrome of Democracy*, 133, 154.

68 Douglas, *Creation of a National Air Force*, 204; Hatch, *Aerodrome of Democracy*, 33, 203. By the end of 1940 the BCATP had produced only seventy-six non-pilot aircrew.

69 Douglas, *Creation of a National Air Force*, 274–5.

70 DHIST, "RCAF Personnel History," 3: 658–65. See also Douglas, *Creation of a National Air Force*, 265.

71 Douglas, *Creation of a National Air Force*, 226–7.

72 DHIST, "BCATP Flying Training: Initial Training Schools, Quality of Instruction," 1–4.

73 Douglas, *Creation of a National Air Force*, 230, 267–9. The RAF also suffered a shortage of instructors, but for a different reason. Opera-

tional demands required that most experienced pilots remain with front-line squadrons: Air Ministry, *Flying Training*, 31. Staffing of RAF Service Flying Training Schools (SFTSs) with fresh graduates began in August 1940: PRO, AHB, "Flying Training," 17.

74 Barris, *Behind the Glory*, 35. Potential instructors with few flying hours to their credit could be sent to a civilian flying club to acquire more experience. As late as October 1938 the pilot certificate test was almost identical to the pre-World War I Fédération aéronautique internationale licence, including the practice of having the examiner observe the test from the ground: 23.

75 Williams, *The Plan*, 36–7, 58, 100; Barris, *Behind the Glory*, 86. When BCATP graduates were posted to the flying schools, they could have accumulated "fewer than 300 hours of flying time": Douglas, *Creation of a National Air Force*, 267.

76 Report of No. 1 Visiting [Standards] Flight, April 1943, quoted in Douglas, *Creation of a National Air Force*, 268.

77 Williams, *The Plan*, 35.

78 In addition, "no standardised, authoritative, technique of flying instruction on monoplane trainers" was followed until after 1941: PRO, AHB, "Flying Training," 43.

79 Hatch, *Aerodrome of Democracy*, 132–3.

80 DHIST, "BCATP Flying Training: The Elementary Flying Training Schools, Wastage and Accidents," 1–2.

81 For a fuller discussion see Douglas, *Creation of a National Air Force*, 269–75.

82 Air Ministry, *Flying Training*, 7, 188.

83 PRO, AHB, "Flying Training," 43, 70; DHIST, "BCATP Flying Training: The Visual Link Trainer," 3–4.

84 PRO, AHB, "Flying Training," 16. The RAF did not give its approval for elementary military flying training by civilian instructors until October 1934 (2), whereas the RCAF's use of civilian instructors began in 1928.

85 Air Ministry, *Flying Training*, 184–7.

86 Williams, *The Plan*, 95.

87 Bott's report is found in RG 24, vol. 5383, HQS 47-6-3, Associate Committee on Aviation Medical Research, "Proceedings of the Executive 29 July 1941," Appendix B, B-3.

88 The official history of the Canadian air force has stated that until the spring of 1943 the RAF was generally pleased with the standard of Canadian-trained pilots: Douglas, *Creation of a National Air Force*, 269–70. Bott's report (see previous footnote) makes it clear, however, that there was the potential for serious problems if OTU failure rates continued.

89 Douglas, *Creation of a National Air Force*, 269–70.

90 Air Ministry, *Flying Training*, 149. In late 1941, 95 per cent of flying training was done during the day, whereas 95 per cent of Bomber Command's operational flying took place at night. It should be noted, however, that when the training syllabi were designed, Bomber Command had planned an offensive based largely on daylight attacks.

91 In January 1941 the RCAF was made aware of the RAF's concerns about insufficient training in instrument flying and the "heavy accident rate at night and in bad weather." It also learned of the steps being taken by the RAF to correct the problem: see RG 24, box 2193, HQS 473–1, Garrod to McKean, 15 January 1941. By February the RCAF had increased the amount of night-flying training given to its students: PRO, AIR 2/8086, f. 1A, Leckie to Garrod, 13 February 1941.

92 PRO, AHB "Flying Training," 62, 66, 71. The RAF was not the only organization to experience problems with training standards between the wars. A senior United States Navy medical officer (MO) commented in 1930 that lack of standardization in American naval flying training created "an impossible situation" for the student, and was contributing to a high rate of wastage in training: Sutton, "Psychology in Aviation," 8.

93 Air Ministry, *Flying Training*, 161–2, 188–91, 194–5, 211–12, 220.

94 Douglas, *Creation of a National Air Force*, 266–8; Hatch, *Aerodrome of Democracy*, 117–18, 154.

95 Signori, "The Arnprior Experiment"; MG 28, Signori transcript; D.R. Hunter, "Aviator Selection," in Wiskoff and Rampton, eds., *Military Personnel Measurement*, 144–5.

96 Douglas, *Creation of a National Air Force*, 267–8.

97 DHIST, "BCATP Flying Training: The Elementary Flying Training Schools, The Grading School Program, Flight Grading Schools," 4–9 (quotation at 5–6); DHIST, "BCATP Flying Training: The Visual Link Trainer," 32–5. The AMP report also credited the flight-grading method of training with improving the quality of BCATP graduates.

98 DHIST, "BCATP Flying Training: The Elementary Flying Training Schools, The Grading School Program, Flight Grading Schools," 6.

99 DHIST, "BCATP Flying Training: The Elementary Flying Training Schools, The Grading School Program, Flight Grading Schools," 1, 6.

100 RG 24, vol. 4947, HQ 895-8/12, Part 1, ff. 168–9, "Syllabus of First Course," paragraphs 51–2.

101 RG 24, vol. 5383, HQS 47-6-3, Associate Committee on Aviation Medical Research, "Proceedings of the Fourth Meeting, 6 June 41," Appendix D, D-3.

102 MG 28, I 161, file 17-9, Humphrey to General Committee [of the CPA], 25 August 1941.

103 MG 28, Williams transcript, 60.

104 There seems to have been some scepticism in the RCAF about scientific training methods. There is evidence of this in a remark by the author of a report to the Associate Committee on Aviation Research, who disparaged "the so-called 'learning process'" in flying training: RG 24, vol. 5383, HQS 47-6-3, Associate Committee on Aviation Medical Research, "Proceedings of the Fourth Meeting, 6 June 41," Appendix D, D-3. Myers encountered the same attitude in the RAF. He found that some senior RAF officers were "offended ... deeply" by his proposal to investigate training problems using scientific methods, because the idea that anyone could conduct scientific experiments in human behaviour was "contrary to their fundamental beliefs": MG 28, Myers transcript, 89.

105 Maycock, *Doctors in the Air*, 77–8.

106 Dr. D.C. Williams, letter to the author, 14 June 1991; Williams, "The Frustrating Fifties"; MG 28, Williams transcript, 60.

107 Air Ministry, *Flying Training*, 237, 250.

108 Douglas, *Creation of a National Air Force*, 284–5; Hatch, *Aerodrome of Democracy*, 145. At No. 6, SFTS, for example, over one-third of the fatal accidents were attributed to "unauthorized low level flying or aerobatics": Barris, *Behind the Glory*, 164.

109 PRO, AHB, "Flying Training," 59–60, 71–2.

110 Douglas, *Creation of a National Air Force*, 293; Air Ministry, *Flying Training*, 2.

111 Air Ministry, *Flying Training*, 221.

112 As Terraine has remarked, it was the CAS who, promoting the "Air Staff point of view," convinced the other two service chiefs to accept the bomber as the war-winning weapon: Terraine, *The Right of the Line*, 290–1, 504–5.

113 Air Ministry, *Flying Training*, 173.

114 This point, which was made in December 1941 by the Inspector General of the RAF, Air Chief Marshal Sir E.R. Ludlow-Hewitt, and endorsed by Arthur Harris and by the AOC No. 5 Group, echoed Tedder's pre-war views: PRO, AIR 33/5, Inspector General's Report No. 218, "Visits to Operational Training Units in Bomber Command," 25 December 1941, 2.

115 Vernon and Parry, *Personnel Selection*, 71–2.

116 The "overall wastage rate in pilot training" in the BCATP was 36 per cent: Hatch, *Aerodrome of Democracy*, 134.

CHAPTER FOUR

1 Feasby, ed., *Official History*, 2: 495.

2 The term currently used by psychologists is Combat Stress Reaction (CSR). It will be discussed in more detail in the next chapter.

3 Dudley Corbett quoted in Robinson, *The Dangerous Sky*, 100–1; Jones, *Flying Vistas*, 200–1.

4 Gibson and Harrison, *Into Thin Air*, 239.

5 Armstrong, *Principles*, 45–6, 56.

6 Two recent works on the subject are Jensen, ed., *Aviation Psychology*, and Gal and Mangelsdorff, eds., *Handbook*.

7 Wilbur, "Aviation," 399.

8 Dudley, "Active Service Flying," 131, 140.

9 Dudley, "Active Service Flying," 131. Army officers, on the other hand, were five times more likely than soldiers of lesser rank to become psychological casualties. This was attributed to the stress of leadership: Bailey, et al., *Medical Department*, vol. 10, *Neuropsychiatry*, 376, 505.

10 Anderson, "Selection," 395.

11 Gilchrist, "Analysis," 401. Gilchrist found that 67 of the 100 aviators he examined were unfit to fly because of mental breakdown. Dudley identified loss of "flying nerve" as responsible for a "very large wastage of pilots": "Active Service Flying," 131. The difficulty of arriving at precise casualty statistics for World War I is discussed in chapter 3.

12 Martin Flack, "Applied Physiology of Aviation," in Anderson, *Medical and Surgical Aspects*, 41; Anderson, *Medical and Surgical Aspects*, 67.

13 In 1914–15, it was reported, 90 per cent of RFC pilot deaths were caused by a deficiency in the pilot (compared to 8 per cent caused by aircraft defect and 2 per cent by enemy action). For 1916–17, however, only 12 per cent of RFC pilot deaths were said to have been caused by a deficiency in the pilot. These figures were cited more often than any others to support the contention that the application of scientific principles to aircrew selection had reduced wastage: Jones, *Flying Vistas*, 192; Birley, "Principles ... Lecture I," 1148, 1151.

14 Birley, "Principles ... Lecture I," 1150. The expression "getting the wind up," meaning "a state of nervous anxiety or fear," came into common use in World War I: *Oxford English Dictionary*, 1989 ed., 20: 388. It did not imply disgrace or misconduct, but was used simply as a factual description of a mental state: Brophy and Partridge, *The Long Trail*, 163. It may have had its origins in the expression "got the wind up his trousers," current in Wellington's army as a synonym for "flustered": Eric Partridge, *A Dictionary of Slang and Unconventional English*, 8th ed. (New York: Macmillan, 1984), 1341.

15 Birley, "Principles ... Lecture I," 1151; Dudley, "Active Service Flying," 132, 134; Gilchrist, "Analysis," 401.

16 Anderson, *Medical and Surgical Aspects*, 88–93; Birley, "Principles ...

Lecture I," 1149; Dudley, "Active Service Flying," 133. The official history of the RFC/RAF in World War I reported that pilots and observers flying in France in "two-seater corps and night-flying squadrons" lasted an average of four months before being rendered unfit for duty by sickness, injury, or death, and three-and-a-half months if they were in "fighter-recce and day-bombing squadrons." Pilots in single-seater fighter squadrons lasted an average of two-and-a-half months: Jones, *War in the Air*, 5: 426.

17 Anderson, *Medical and Surgical Aspects*, 63–4; Dudley, "Active Service Flying," 133–4; Gotch, "Aero-Neuroses," 128–33.

18 Birley, "Principles ... Lecture I," 1151.

19 Bailey, et al., *Medical Department*, vol. 10, *Neuropsychiatry*, 506–10, 520.

20 Bailey, et al., *Medical Department*, vol. 10, *Neuropsychiatry*, 506.

21 Bailey, et al., *Medical Department*, vol. 10, *Neuropsychiatry*, 511. The Granville Canadian Hospital at Ramsgate was an exception, for it was reported to be able to return "upward of 60 percent" of its patients to the front: 511.

22 Birley rose to the rank of lieutenant-colonel in the RAF and was twice mentioned in despatches. After the war he served as a consulting physician to the RAF and, in 1920–22, as a member of the War Office Committee on Shell Shock. He died in 1934 at age fifty: *Who Was Who* (London: Adam & Charles Black, 1947), 3: 115–16.

23 Birley, "Principles ... Lecture I," 1149.

24 Strange, *Recollections*, 169.

25 See, for example, Strange, *Recollections*, 79; Lewis, *Wings over the Somme*, 81, 140, 175; Wortley, *Letters*, 168.

26 Taylor, *Sopwith Scout 7309*, 107.

27 Birley, "Principles ... Lecture I," 1149.

28 Insall, *Observer*, 69.

29 Villars, *Notes*, 97, 146–7, 169, 200; Vee, *Flying Minnows*, 239; Kinney, *I Flew a Camel*, 88.

30 Macmillan, *Into the Blue*, 47.

31 Raleigh, *War in the Air*, 1: 438.

32 Flack, who rose to the rank of group captain, died in 1934 at age fifty-two: *Who Was Who*, 3: 454.

33 Dockeray and Isaacs, "Psychological Research," 127–33; Armstrong, *Principles*, 27–8; Robinson, *The Dangerous Sky*, 84–5; Flack, "Applied Physiology," 41, 55–8.

34 RG 24, vol. 5383, HQS 47-6-3, vol. 1, Associate Committee on Aviation Medical Research, "Proceedings of the Executive 29 July 1941," Appendix G. This report stated that even though the aircrew-selection "standards adopted by the RAF and RCAF have been based on experience ... it is still felt that a great many of them are somewhat arbi-

trary" and that "considerable investigative work is indicated." See also, in the same volume, "Proceedings of the Associate Committee on Aviation Medical Research 18 September 1941," Appendix A, and Armstrong, *Principles*, 46–9.

35 RG 24, vol. 5383, HQS 47-6-3, vol. 3, United States National Research Council, Division of Medical Science Committee, Aviation Medical Reports, Report 43, 3 April 1942.

36 Birley, "Lecture," 781, 783–4.

37 E.J. Dearnaley and P.B. Warr, "Preface," in Dearnaley and Warr, eds., *Aircrew Stress*, v. Bartlett died in 1969 at age eighty-three: *Who Was Who*, 6: 65.

38 Bartlett, *Psychology*, 11, 35–6.

39 Conversion hysteria refers to a condition where the patient has the symptoms of a physical injury or disability, but the causes are entirely psychological: Bartlett, *Psychology*, 180–1.

40 Bartlett, *Psychology*, 187.

41 Bartlett, *Psychology*, 191.

42 Bartlett, *Psychology*, 197.

43 Bartlett, *Psychology*, 203, 207–8.

44 Bartlett, *Psychology*, 204, 210–12.

45 Dunlap, "Psychological Research," 96; Thorndike, "Scientific Personnel Work," 55, 57.

46 Neuropsychiatry is defined as "clinical neurology combined with psychiatry": *Webster's Dictionary*, 2nd ed., 1952.

47 Rees, "Three Years," 1.

48 Bailey, et al., *Medical Department*, vol. 10, *Neuropsychiatry*, 369, 399, 497, 501, 508–9, 513. In the British Army it was found that although officers made up about 3 per cent of the fighting men at the front, they accounted for 17 per cent of the "war neurosis" cases studied up to April 1917. It was estimated that not less than 1 per cent of all British Army soldiers who served outside of Britain became casualties as a result of a war neurosis: 505, 507. The official medical history of the British war effort stated that there was "little information regarding the wastage due to nervous disorders": Mitchell and Smith, *History*, 115. In a sample of just over one million British casualties after 1916, "functional diseases of the nervous system" and "mental diseases" account for 2 per cent of all cases. In the same sample, influenza (9 per cent of all casualties) stood first among the diseases. Wounds accounted for 20 per cent of the total casualty figure: 278, 285.

49 Bailey et al., *Medical Department*, vol. 10, *Neuropsychiatry*, 369.

50 In the 1930s there was considerable debate over whose views should govern the treatment of nervous disorders. The competing disciplines included psychology, psychiatry, neurology, and neuropsychiatry:

Symonds, "The Neurological Approach to Mental Disorder," in Symonds, *Studies*, 246–9 (this article first appeared in the *Proceedings of the Royal Society of Medicine* in 1941). Within the medical profession, both neurologists and psychiatrists were involved in the treatment of RAF psychological casualties during World War II. Most of these specialists called themselves neuropsychiatrists. Occasionally, however, there was confusion about how a specialist should be described. Symonds, for example, was at one point referred to as a consultant in neurology, but later was called a consultant in neuropsychiatry: Symonds and Williams, "Psychological Disorders ... Section 2," 22–4.

51 Air Ministry, "Notes," 9, 16, 18. Bartlett's *Psychology* and MacCurdy's *War Neuroses* (which Bartlett cited frequently in his book) were two of six works recommended to MOs in the Air Ministry's list of suggested readings in the "Notes": 20.

52 Air Ministry, "Notes," 4–7, 16–17.

53 Freud hypothesized that the demands of the id (the "lower," "instinctive" drives) and the demands of the superego (moral demands) are mediated by the ego, which establishes compromises between the two sets of demands. These compromises are what is observed as behaviour in humans: Fancher, *Pioneers*, 380–2.

54 Air Ministry, "Notes," 2, 4–5, 8–9, 15–16, 18–19.

55 Air Ministry, "Notes," 7, 18.

56 Air Ministry, "Notes," 7. The "result" the pamphlet was referring to was the patient's reaction to treatment, but the implication was obvious: airmen with the "right stuff" would continue to fly, those without it would not.

57 Leading British psychologists of the day, such as Cyril Burt and Charles Spearman, were staunch supporters of the hereditarian views of Sir Francis Galton: Fancher, *The Intelligence Men*, 84–6, 170; Fancher, *Pioneers*, 213, 219, 237. They were also influenced by American psychologists who concluded, based on mass intelligence testing conducted during World War I, that there was a strong positive correlation between high intelligence and good emotional control: Gould, *The Mismeasure of Man*, 160–1. Recent scholarship has shown that while IQ test results frequently correlate well with quickness to learn or performance during job training, they often bear little or no relation to how a job is performed after training: Fancher, *The Intelligence Men*, 228.

58 See, for example, Buckle, "Prevention," 125. A psychiatrist in the Royal Australian Air Force, Buckle drew on psychoanalytic concepts to describe an alleged predisposition toward psychological disorders among some airmen. He said that predisposition could be explained by "the amount of infantile dependence [in the airman's personality] and the ease with which he can regress to earlier emotional levels."

59 Gillespie, *Psychological Effects of War*, 180, 221; Air Ministry, "Notes," 8.

60 *Who's Who, 1976*, 2321–2. On Birley's death Symonds became a civilian consultant in neurology to the RAF. he had served as a doctor with the Royal Army Medical Corps during World War I, returning to private practice after the war. He was promoted air vice-marshal in early 1945. Symonds believed that "inherited disposition plays an important part" in psychological disorders. However, while he accepted that Freud had "shown the meaning of individual symptoms," he felt that "the analytic method on the whole has been responsible for as much harm as good in the treatment of affective disorder": Symonds, *Studies*, 1, 16, 21, 243–5.

61 *Who's Who 1976*, 2561–2.

62 D.D. Reid, "Historical Background to Wartime Research in Psychology in the RAF," in Dearnaley and Warr, eds., *Aircrew Stress*, 1; *Who's Who 1976*, 1983. All three men went on to distinguished medical careers after the war.

63 A flying personnel MO was a medical doctor who had taken flying training and qualified to at least wings standard. His main function was to advise the Commander-in-Chief (C-in-C) and the Principal Medical Officer of each Command on aircrew health and other matters related to aviation medicine: Maycock, *Doctors*, 92. One author called the squadron and station MOs of the RAF the "G.P.s" of aviation medicine, whereas the flying personnel MO was the "specialist": S.C. Rexford-Welch, "The RAF Medical Services," in MacNalty and Mellor, eds., *Medical Services*, 246. For a detailed description of the duties of Bomber Command's flying personnel MO, see Rexford-Welch, ed., *Royal Air Force Medical Services*, 2: 137–45.

64 See Corner, "Flying Duties"; Gillespie, "Report"; Grindley, "Note on 'Flying Stress'"; Symonds, "Report."

65 Symonds, "Memorandum," 3. The complete list comprised anxiety, depression, elation, fatigue syndrome, hysteria, obsessional, schizophrenia, organic acute, and organic chronic.

66 Neurasthenia is defined as a "debility of the nerves causing fatigue and listlessness": *Oxford Concise Dictionary*, 1976 ed., 733.

67 Symonds, "Memorandum," 3. Air Ministry, "Notes," used the term "neurasthenia" and did not refer to "fatigue syndrome": 10.

68 Symonds and Williams, "Psychological Disorders ... Section 1" 1, 5, 7, 8–11.

69 Symonds and Williams, "Psychological Disorders ... Section 2," 2.

70 Symonds developed this concept further in lectures delivered at Harvard in March 1943 and to the Royal College of Physicians the following May, in which he advanced a physiological explanation for the ways in which different types of people handled fear: C.P. Sy-

monds, "The Human Response to Flying Stress," in Air Ministry, *Psychological Disorders*, 110–11.

71 Symonds and Williams, "Psychological Disorders ... Section 2," 4, 6.

72 Symonds and Williams, "Psychological Disorders ... Section 2," 10.

73 Symonds and Williams, "Psychological Disorders ... Section 2," 7.

74 Symonds and Williams, "Psychological Disorders ... Section 2," 17.

75 The military was influenced by research which suggested that as much as one-quarter of the population of the British Isles were "dull and backward," "feeble-minded," "imbeciles," or "idiots." In addition, two-fifths of this group (one-tenth of the total British population) qualified as "psychopathic": Rees, "Three Years," 2–3.

76 Symonds and Williams, "Psychological Disorders ... Section 2," 5.

77 Symonds and Williams, "Psychological Disorders ... Section 2," 5.

78 See, for example, Stafford-Clark "Morale"; Ironside and Batchelor, *Aviation Neuro-Psychiatry*. Stafford-Clark was a Bomber Command MO for over four years. After the war he became a leading British psychiatrist and published extensively: Stafford-Clark, "Morale," 14; *Who's Who 1976*, 2245–6. Ironside reached the rank of air commodore and was an RAF consultant in neurology and neuropsychiatry during the war: *Who Was Who*, 6: 582; Ironside and Batchelor, *Aviation Neuro-Psychiatry*, i. Batchelor was a squadron leader MO and oversaw the neuropsychiatric division of an RAF general hospital during the war. After the war he became a prominent British psychiatrist: *Who's Who 1980*, 154.

79 Moran, *Anatomy of Courage*, xv–xvi, 22–3 (quotation at 22–3).

80 Ironside and Batchelor, *Aviation Neuro-Psychiatry*, particularly 10–12, 14–16, 34 (quotation at 34).

81 Williams, "Predisposition," 192.

82 Air Ministry, *Psychological Disorders*, i.

83 The most important completed study used a control group and involved 100 aircrew admitted to hospital for "acute head or spinal injuries": Williams, "Predisposition," 187. A study of 1,000 bomber pilots was begun in 1944, but the war ended before any definitive results could be seen: Reid, "Historical Background," 4. A Canadian study involved the psychiatric assessment of 1,000 aircrew trainees at ITS, followed by a comparison of the psychiatrists' predictions and student performance results up to SFTS level. The study concluded that psychiatric assessment was not as effective as Link Trainer scores in predicting success in training. Neither psychiatric assessment nor Link trainer scores had predictive values for success on operations: Department of National Defence (DND), "Follow-up Study," 1–2, 6.

84 Shabtai Noy, "Combat Stress Reactions," in Gal and Mangelsdorff, eds., *Handbook*, 514.

85 See, for example, Ironside and Batchelor, *Aviation Neuro-Psychiatry*, v. Ironside and Batchelor admitted that their book suffered from an "absence of references" because it had been written while the authors were on active duty overseas. They hoped to compensate for its lack of comparative evidence by basing their conclusions on the validity of personal experience. Some aviation psychiatrists still use the expression today to defend small data sets or statistically inconclusive results: Wenek interview with the author.

86 Chalke, "Psychiatric Screening," 282–4, 290. While most of Chalke's complaints were directed at research done on the psychiatric screening of recruits, they were equally applicable to the identification and diagnosis of psychological disorders. For a British Army MO's criticisms see L'Etang, "A Criticism of Military Psychiatry," 192–5.

87 Quote in Robinson, *The Dangerous Sky*, 183–4. The same opinion was also held by some doctors in the British Army: Rees, "Three Years," 4.

88 Grinker and Spiegel, *Men Under Stress*, vii.

89 Ironside and Batchelor, *Aviation Neuro-Psychiatry*, 32.

90 See, for example, Anderson, *Medical and Surgical Aspects*, 97; Birley, "Principles ... Lecture I," 1148; Rees, "Three Years," 2.

91 Reid, "Prognosis," 241, 243.

92 Hill and Williams, "Investigation," 1, 11.

93 Hitchins, *Air Board*, 94, 187, 239, 339, 362. To keep up with the expansion of the RCAF in the immediate pre-war period, the number of Royal Canadian Army Medical Corps (RCAMC) doctors seconded to the air force grew to a dozen.

94 Feasby, ed., *Official History*, 1: 338–43, 354.

95 Feasby, ed., *Official History*, 1: 428.

96 Feasby, ed., *Official History*, 1: 341–2, 353–4, 356–7.

97 Hitchins, *Air Board*, 375.

98 RG 24, vol. 4947, HQ 895-8/12, Part 1, Tice to SOMS (Air), 31 January 1940.

99 Armstrong, *Principles*, 142, 159.

100 Mashburn, "Some Interesting Psychological Factors," 113–14, 120, 123, 126.

101 Symonds was also quite critical of Armstrong's work in aviation psychology: Symonds, "Memorandum," 2–3.

102 Diringshofen, *Medical Guide*, 97–102; G.E. Subbotnik, "Investigation of the Nervous System and the Psychic State," in Pavlov Institute, *Fundamentals*, 292–3, 295–8.

103 RG 24, vol. 4947, HQ 895-8/12, Part 1, f. 169, "Syllabus of First Course," paragraph 52.

104 Feasby, ed., *Official History*, 1: 456.

105 RG 24, vol. 4947, HQ 895-8/12, Part 2, Ryan to Mathewson, 13 February 1941.

106 RG 24, vol. 4947, HQ 895-8/12, Part 2, CO 1 Manning Depot to Director of Medical Services (Analysis of 7th Course Exam Questions), 6 August 1941.

107 RG 24, vol. 5383, HQC 47-3-1, RAF Component France, Technical Medical Instruction No. 1, 14 February 1940 (sent to Canada in May 1940 in Report of Tour of Principal Medical Officer, RCAF [K.A. Hunter]).

108 RG 24, vol. 5383, HQC 47-3-1, Report No. 2 of the Principal Medical Officer of the RCAF [Overseas], 20 March 1940, f. 42.

109 Feasby, ed., *Official History*, 1: 360.

110 RG 24, vol. 5383, HQC 47-3-1, Report of the Principal Medical Officer of the RCAF Overseas, 19 February 1940.

111 Feasby, ed., *Official History*, 1: 361; Haycock, *Hughes*, 313. The Canadian air force was able to achieve only a small measure of autonomy in the medical treatment of its flyers in Britain: the establishment of large sick-quarters at several RCAF stations; treatment centres shared with the RCAMC; access to RCAMC general hospitals; and the posting of Canadian medical staff to RAF hospitals, in particular the hospitals of No. 6 Group: Feasby, ed., *Official History*, 1: 361–6, 375, 544.

112 RG 24, vol. 5383, HQS 47-6-3, "Proceedings of the Associate Committee on Aviation Medical Research, 18 September 1941," Addenda to Minutes, Report of Wing Commander J.W. Tice's visit to England.

113 Feasby, ed., *Official History*, 1: 354; 2: 96–7.

114 Feasby, ed., *Official History*, 2: 366–9. The RCAF Central Medical Board processed 6,312 cases to the end of May 1945.

115 Feasby, ed., *Official History*, 1: 369.

116 Feasby, ed., *Official History*, 1: 467. One of the main differences was that Canadians were "accustomed" to having their illnesses and treatments explained to them: 1: 369.

117 Gillespie, *Psychological Effects of War*, 19, 32.

118 Ironside and Batchelor, *Aviation Neuro-Psychiatry*, 10.

119 Thomas P. O'Hearn, Jr., "Psychotherapy and Behaviour Change," in Gal and Mangelsdorff, eds., *Handbook*, 609.

120 Noy, "Combat Stress Reactions," 509–10.

121 The volume in Britain's official medical history subtitled "The Principal Medical Lessons of the Second World War" commented that "'almost all the lessons'" of World War I had been forgotten between the wars, and that "'one of the most serious deficiencies on the part of psychiatry as a whole [in World War II] was its emphasis on the individual, almost as an isolated unit independent of group dynamics, and its relative neglect of "social psychiatry'": quoted in

R.H. Ahrenfeldt, 'Military Psychiatry," in MacNalty and Mellor, eds., *Medical Services*, 176–7.

CHAPTER FIVE

1 PRO, AIR 33/5, E.R. Ludlow-Hewitt, Inspector General's Report No. 271, "Visits to Air Crew Disposal and Suspendair Centres ...," 26 December 1942, 2. The term LMF was not restricted to the air force. Terraine has recorded that Major-General H. Essame, like many World War I army veterans, thought "'that there was a lack of moral fibre in the younger generation of soldiers'": Terraine, *The Right of the Line*, 640.

2 Hastings, *Bomber Command*, 249–50.

3 Terraine, *The Right of the Line*, 521, 532; McCarthy, "Aircrew," 96–7.

4 Terraine, *The Right of the Line*, 528, 535.

5 McCarthy, "Aircrew," 87.

6 Longmate, *The Bombers*, 188; Schurman interview with the author (Professor Schurman flew with the RCAF's No. 6 Group).

7 Hastings, *Bomber Command*, 245.

8 Rexford-Welch, ed., *Royal Air Force Medical Services*, 2: 122–37.

9 *Oxford English Dictionary*, 2nd ed. (1989), 8: 1069.

10 McCarthy, "Aircrew," 88 n.2.

11 Symonds and Williams, "Psychological Disorders ... Section 2," 17.

12 Ironside and Batchelor, *Aviation Neuro-Psychiatry*, 17–20 (quotation at 10).

13 McCarthy, "Aircrew," 87–8.

14 Terraine, *The Right of the Line*, 533.

15 My emphasis.

16 PRO, AIR 2/8591, s.61141/s.7.C(1), "Memorandum on the Disposal of Members of Air Crews who Forfeit the Confidence of their Commanding Officers," 19 September 1941, 2–3.

17 Rexford-Welch, ed., *Royal Air Force Medical Services*, 2: 133. Most of this volume's comments on flying stress are identical to those found in the No. 3 Group medical history reports written from December 1941 to June 1942: PRO, AIR 14/2821. Perhaps this should not be surprising, for two prolific writers on flying stress, D.D. Reid and D. Stafford-Clark, were No. 3 Group SMOs.

18 Rexford-Welch, ed., *Royal Air Force Medical Services*, 2: 133.

19 PRO, "Memorandum," 3–4; Hastings, *Bomber Command*, 249; McCarthy, "Aircrew," 88–9, 96; PRO, AIR 2/8592, "Minutes of a meeting held on the 20th Oct 1944, to discuss the 'w' procedure." The quotation is from Hastings.

20 See, for example, Messenger, *"Bomber" Harris*, 205–7.

21 Rexford-Welch, ed., *Royal Air Force Medical Services*, 2: 133. Symonds found that the attitudes of "General Duties" [aircrew] officers toward LMF varied: Symonds, *Studies*, 19.

22 Hastings, *Bomber Command*, 243.

23 PRO, Inspector General's Report No. 271, 2; Appendix I, 1; Attachment B.1 to List A.1, 1; List A.2, 3.

24 Webster and Frankland, *Strategic Air Offensive against Germany*, 4: 401.

25 Reid, "Study," 10, 14. Reid did not accept Symond's definition of flying stress as a cause of neurosis, but saw it as an "occupational disease": 1.

26 Symonds and Williams, "Clinical and Statistical Study," 33; Symonds, *Studies*, 19.

27 Symonds and Williams, "Clinical and Statistical Study," 34.

28 Ironside and Batchelor, *Aviation Neuro-Psychiatry*, 74.

29 Symonds stated that "Our objective was the conservation of military manpower": Symonds, *Studies*, 17.

30 Symonds and Williams, "Personal Investigation," 53; Corner, "Flying Duties," 1–2; Gillespie, "Report," 1; Rexford-Welch, ed., *Royal Air Force Medical Services*, 2: 42–5, 127, 131–2.

31 Symonds and Williams, "Psychological Disorders ... Section 2," 4, 8–9. OTU losses were sometimes catastrophic. For example, on one of the first thousand-bomber raids, when OTU crews were used to bring the number of attacking aircraft up to the magic number of 1,000, 91 Group (an OTU Group) lost twenty-one of the twenty-four aircraft it dispatched: Messenger, *"Bomber" Harris*, 81. Such severe losses could contribute to higher levels of flying stress in the OTUs than were found in operational squadrons.

32 Symonds and Williams, "Psychological Disorders ... Section 2," 10–12. For the rendering of NYDN as "Not yet diagnosed nervous" see Privy Council Office, "Report," 95. The official medical history of the RAF has "Not yet diagnosed-Neuropsychiatric?": Rexford-Welch, ed., *Royal Air Force Medical Services*, 2: 127.

33 Symonds and Williams, "Psychological Disorders ... Section 2," 15–17, 19.

34 Symonds and Williams, "Psychological Disorders ... Section 2," 14–15.

35 Symonds and Williams, "Probability of Return to Full Flying," 1.

36 Symonds and Williams, "Investigation," 6.

37 Symonds and Williams, "Clinical and Statistical Study," 34.

38 Hill and Williams, "Investigation," 1.

39 Symonds and Williams, "Investigation," 6. Emphasis in original.

40 *Bomber Command Quarterly Review*, 14.

41 Rachman, *Fear and Courage*, 2, 7–8, 43–4.

42 Quoted in Rachman, *Fear and Courage*, 50.

43 A useful summary of the literature on primary groups may be found in Kellet, *Combat Motivation*, 41–7, 97–104.

44 Rachman, *Fear and Courage*, 52–3.

45 Rachman, *Fear and Courage*, 65.

46 Rachman, *Fear and Courage*, 58–9.

47 Rachman, *Fear and Courage*, 56–7.

48 Hastings, Wright, and Glueck, "Psychiatric Experiences of the Eighth Air Force," 125–6, 153–4.

49 Rachman, *Fear and Courage*, 57; Hastings, Wright, and Glueck, "Psychiatric Experiences of the Eighth Air Force," 4.

50 Emphasis in original.

51 Hastings, Wright, and Glueck, "Psychiatric Experiences of the Eighth Air Force," 4, 17, 58–9, 154, 158, 207.

52 Hastings, Wright, and Glueck, "Psychiatric Experiences of the Eighth Air Force," 160–1, 167–70. This scheme was similar to one used by Bomber Command shortly after the outbreak of war.

53 Hastings, Wright, and Glueck, "Psychiatric Experiences of the Eighth Air Force," 198–200, 208.

54 Most of the recent CSR research has been conduced on army units; however, there are many areas where the results are generalizable to air combat.

55 Shabtai Noy, "Combat Stress Reactions," in Gal and Mangelsdorff, eds., *Handbook*, 508.

56 Noy, "Combat Stress Reactions," 510–13. The correlation between numbers of wounded and numbers of CSR casualties was also noted by the RAF in World War II. See Symonds and Williams, "Psychological Disorders ... Section 2," 8; Reid, "Study," 10.

57 Noy, "Combat Stress Reactions," 513–15. The idea that some people are more resistant to stress than others has been advanced by other authors. See, for example, Allred and Smith, "The Hardy Personality," 257–66.

58 Noy, "Combat Stress Reactions," 517. RAF MOs also noted that "When a commanding officer states he has lost his confidence in the reliability of a member of aircrew he stigmatises that individual's efficiency": Ironside and Batchelor, *Aviation Neuro-Psychiatry*, 19.

59 Dudley, "Active Service Flying," 134.

60 Noy, "Combat Stress Reactions," 520–1.

61 See, for example, Rexford-Welch, ed., *Royal Air Force Medical Services*, 2: 129, 132; Grinker and Spiegel, *Men under Stress*, 26; Hastings, *Bomber Command*, 245; Longmate, *The Bombers*, 186.

62 Frederick J. Manning, "Morale, Cohesion and Esprit de Corps," in Gal and Mangelsdorff, eds., *Handbook*, 464.

63 Hastings, *Bomber Command*, 247–8, 252.

64 In addition to the ways already mentioned in this chapter, Hastings, *Bomber Command*, 248 has pointed to the practice of deliberately fouling the magnetos while running-up the engine, and Longmate, *The Bombers*, 184 to cases of tampering with gun-turret hydraulic systems. Air Vice-Marshal Bennett described the number of bombs jettisoned prematurely during the Battle of Berlin as "'enormous'": Webster and Frankland, *Strategic Air Offensive against Germany*, 2: 195–6. Harris's policy of having tour lengths measured by the number of successful sorties, and where possible confirmed by photos taken at bomb release, was designed to discourage "fringe merchants," "boomerangs," those who jettisoned their bombs, and aircrews who deliberately stretched out their trips (at a time when a tour length was set at 200 hours' flying time): Messenger, *"Bomber" Harris*, 90; Terraine, *The Right of the Line*, 524.

65 PRO, AHB, "Flying Training," 14.

66 Apparently this innovation was "little used": Rexford-Welch, ed., *Royal Air Force Medical Services*, 2: 127.

67 Symonds and Williams, "Psychological Disorders ... Section 2," 19.

68 PRO, AHB, "Flying Training," 20.

69 Stafford-Clark, "Morale," 18.

70 This effect had been noted in early 1940 by flying personnel MOs in Fighter and Coastal Commands. Being on standby could generate 25 per cent to 80 per cent as much stress as actual combat flying, and created so much fatigue that there were "several cases of pilots falling asleep in the air": Corner, "Flying Duties," 1–2.

71 See also Stafford-Clark, "Morale," 10.

72 Symonds and Williams, "Personal Investigation," 37, 43–6, 51–3.

73 Symonds and Williams, "Personal Investigation," 57. The usefulness of military discipline is still the subject of much debate. A valuable summary of the discussion can be found in Shalit, *Psychology*, 122–47. Shalit has agreed with S.L.A. Marshall's conclusions that pre-battle discipline does not predict battle discipline, and that good discipline is a function of high morale, not its precursor.

74 Symonds and Williams, "Psychological Disorders ... Section 2," 15–16, 21 (quotation at 15).

75 Buckle, "Prevention of Psychiatric Disorders," 125.

76 PRO, AIR 14/290, BC/C. 23068/P., Air Officer Administration Bomber Command to AOCS 1, 2, 3, 4, 5, 6 Groups and Advanced Air Striking Force, 28 October 1939, and minutes 1–18; BC/S.24096/P., C-in-C Bomber Command to Under-Secretary of State, 18 June 1940. Portal's position was supported by Harris in January 1941 when Harris was Deputy Chief of the Air Staff: Saward, *"Bomber" Harris*, 90–1.

77 No. 6 (Training) Group was renumbered to No. 91 Group with the

formation of No. 6 (RCAF) Group: Terraine, *The Right of the Line*, 462.
PRO, AIR, 14/1954, Smyth-Pigott to Air Commodore MacNeece Foster,
AOC 6 Group, 23 June 1941. Group Captain Ruscombe Smyth-Pigott
had won the Distinguished Service Order in World War I and served
with the RAF for a short time after the war. He was recalled to service
at the beginning of World War II, and commanded RAF Station Per-
shore in 1941: Webster and Frankland, *Strategic Air Offensive against
Germany*, 1: 256 n.7; *Who's Who 1948*, 2582; *RAF List*, January 1939.

78 Harris was prepared to lose sixty aircraft: Messenger, *"Bomber"
Harris*, 77.

79 Quoted in Longmate, *The Bombers*, 221.

80 Longmate, *The Bombers*, 217, 221–3. After the first thousand-plane
raid, Harris forbade group commanders from flying on operational
missions, on the grounds that they knew too many secrets about
Bomber Command that would be of use to the enemy if they were
captured: Messenger, *"Bomber" Harris*, 202.

81 Messenger, *"Bomber" Harris*, 202.

82 See also Hastings, *Bomber Command*, 159–60; Schurman interview with
the author. Schurman said that COS who attempted to minimize
losses "one way or another" were respected by their squadrons.

83 Symonds and Williams, "Personal Investigation," 32, 53.

84 Symonds and Williams, "Personal Investigation," 53–4.

85 Terraine, *The Right of the Line*, 534–5. Wing Commander J. Lawson
worked in the Air Ministry Personnel Department and "was intimately concerned with the handling of [LMF] cases": 532. According to
Terraine, Bomber Command strength in June 1944 was 155,510; total
RAF strength in October 1944 was 1,171,421: 535.

86 Verrier, *The Bomber Offensive*, 194, has rightly called LMF casualties
"small indeed," but has neglected, as have other commentators, to
consider other losses attributable to psychological causes.

87 This figure is derived from the annual reports prepared by RAF
neuropsychiatrists, as summarized in Symonds and Williams, "Occurrence of Neurosis ... 1944 and 1945," 178. The annual totals were
as follows: 2,503 (1942–43), 2,989 (1943–44), and 2,910 (1944–45). It
should be recognized that the 8,402 included some individuals who
were chronic neurotics or even psychotics. No attempt to distinguish
between those who had some sort of mental illness before joining the
air force and those whose disease was a result of service conditions
was made. Some indication of how many people entered the RAF
with an existing mental illness may be obtained from the fact that 6
per cent to 7 per cent of the cases referred to neuropsychiatrists had
suffered a breakdown before beginning flying training: Symonds and
Williams, "Occurrence of Neurosis ... 1944 and 1945," 180. Because
the selection system could not weed out such cases, and because they

imposed a drain on the system, no attempt has been made to exclude them from the present calculations.

88 Symonds and Williams, "Occurrence of Neurosis ... 1944 and 1945," 178. The medial reports indicated that 416 aircrew were found to be "lacking confidence" in 1942–43, 307 in 1943–44, and 306 in 1944–45. The categories "not having a neurosis but unsuitable for flying duties" and "considered normal and returned to duty" were first reported in 1944–45.

89 Rexford-Welch, ed., *Royal Air Force Medical Services*, 2: 128, claimed that "monthly statistical records" on "all cases" of "flying stress" were kept at Bomber Command headquarters from 1 April 1940 and were "supplied to the Consultant in Neuropsychiatry," meaning Symonds. But although Symonds wrote numerous reports, often based on returns from RAF neuropsychiatrists and on individual case studies, I have not discovered any report that makes use of these monthly records.

90 This study (Symonds and Williams, "Investigation," 1–2, 10) is the only one I have found that examined the treatment of psychological disorders by unit MOs. "[No] great reliability [was] claimed for the data" by Symonds and Williams: 1–2. The study presented statistics on 286 aircrew in the twenty weeks beginning 15 May 1944 who saw their MOs "for complaints which have no physical basis, or who having no physical disability state[d] that they [felt] unable to carry on with flying duties." It reported that MOs treated about 70 per cent of the cases. Of the remainder, who were not treated on station, 30 per cent (9 per cent of the total population in the study) were diagnosed LMF and dealt with by the executive branch, 60 per cent (18 per cent of the total population in the study) had a "bad or hopeless prognosis," and 38 per cent were "under training and considered unsuitable for aircrew duties." No explanation was given as to why the percentage breakdown for the aircrew not treated on station added up to more than 100 per cent, but in other reports numbers in excess of 100 per cent resulted from individuals being assigned to more than one diagnostic category.

There is no information with respect to how many of the cases treated by MOs at the unit level were eventually referred to specialists; a conservative estimate might be 30 per cent (21 per cent of the total population of the study). In addition, some of the 9 per cent of the total population who were designated LMF and the 18 per cent of the total population who were given a "bad or hopeless prognosis" would undoubtedly have been referred to specialists. I suggest, therefore, that of all the neurosis cases seen by RAF unit MOs, about 30 per cent were examined by specialists on referral.

91 Symonds and Williams, "Investigation," 1, 10.

92 The 10,000-a-year figure is a reasonable one, I believe, when we consider that the RAF maintained four hospitals to provide, in the words of the 1947 Privy Council report on psychologists and psychiatrists in the armed services, "intense and prolonged psychiatric treatment," as well as fourteen NYDN Centres (twelve in the United Kingdom and two overseas). The report also stated that during the war the RAF's MOS were expected to care for only "the mildest cases"; all others were to go to the nearest NYDN Centre. The only statistics I have found relating to hospital or NYDN Centre treatment come from RAF Hospital Matlock, which treated both aircrew and ground personnel, but not officers. Its annual admission rate was about 1,200: Privy Council Office, "Report," 63–5, 95.

93 Estimates of the proportion of RAF "neurosis" cases in Bomber Command vary. My figure of 40 per cent is the average of the annual rates of "recognised neurosis" given in two FPRC reports covering the years 1942–43, 1943–44, and 1944–45: Symonds and Williams, "Clinical and Statistical Study," 39; Symonds and Williams, "Occurrence of Neurosis ... 1944 and 1945," 178. The figure of 40 per cent is also given in Privy Council Office, "Report," 65–6.

94 I have based my calculations on the total number of flying personnel who were in Bomber Command from 1 January 1943 to 31 December 1943, as this gives a truer picture of the population within which the psychological casualties occurred. In 1943, 34,269 flying personnel served in the OTUs of Bomber Command, and 44,985 in operational units, for a total of 79,254. The large number of men who served in the Command relative to its established strength (about 8,000) was primarily due to the high casualty rates on bomber operations: Rexford-Welch, ed., *Royal Air Force Medical Services*, 2: 122.

95 Symonds and Williams, "Psychological Disorders ... Section 2," 5. The fact that the pre-1941 total only included those aircrew admitted to hospital or dealt with administratively does not invalidate the comparison. MOS at that time had "little or no experience" with psychological problems, and they were not closely involved in treatment before 1941; therefore, many cases had to be hospitalized to receive care. Later, most cases were treated at the unit level, by MOS without recourse to more specialized resources: Rexford-Welch, ed., *Royal Air Force Medical Services*, 2: 124, 127. The figure in per cent man-years (which approximates percentage of average strength) from 10 February 1942 to 9 February 1943 was 12.1 for night bombers: Symonds and Williams, "Clinical and Statistical Study," 141, 145–6. The figure for 1943–44 was 8.5: Williams, "Note," 2.

96 The 10 per cent does not, of course, include airmen who were killed during training or in combat before their illnesses could be identified.

Symonds and Williams stated that the statistics did not include "those who because of psychological disorders [were] handicapped in their duties, and though inefficient [did] not reach the medical branch": Symonds and Williams, "Occurrence of Neurosis ... 1943 and 1944," 173. The statistics also did not include, first, those found "unsuitable on constitutional grounds during training" and "removed without medical advice," and second, those "lacking in confidence" and "removed without medical advice," and second, those "lacking in confidence" and "disposed of by the executive without the opinion of a neuro-psychiatric specialist": Symonds and Williams, "Statistical Survey," 117–18. There was, moreover, a great deal of peer pressure among aircrew not to report sick, especially if it meant breaking up a crew, and accordingly some aircrew were given what was called "special handling" that is, they were treated with compassion to get them through a tour. Even the RAF's official medical history admitted that the General Duties [aircrew] Branch attached a slur to flying stress or saw it as synonymous with LMF. For this reason, "many experienced aircrew" refused to believe in flying stress: Rexford-Welch, ed., *Royal Air Force Medical Services*, 2: 126.

97 As Symonds acknowledged, statistics were available only on those aviators who returned from operations. Owing to Bomber Command's high casualty rate and the nature of its operations, many were lost without explanation: Symonds and Williams, "Personal Investigation," 34. Of 22,330 operational casualties in 1943–44, 14,933, or two-thirds, were missing and presumed dead: Webster and Frankland, *Strategic Air Offensive against Germany*, 4: 440.

98 Bomber Command incurred 19,096 operational and non-operational casualties during the year 3 September 1942–2 September 1943. For the following year the figure was 25,896. The average of these two numbers is 22,496, which for our purposes has been rounded to 20,000: Webster and Frankland, *Strategic Air Offensive against Germany*, 4: 440.

99 Stokes and Kite, *Flight Stress*, 217; Chappelow, "Human Factors in Aviation," 223.

100 The crew size of "the principal aircraft in Bomber Command" varied from two in the Mosquito to seven, or more, in the "heavies" such as the Halifax and Lancaster: Webster and Frankland, *Strategic Air Offensive against Germany*, 4: 447. Two-thirds of the Command's sorties were flown by heavy bombers: Middlebrook and Everitt, *The Bomber Command War Diaries*, 707.

101 Symonds and Williams, "Probability of Return to Full Flying," 1; Reid, "Prognosis," 237; Symonds and Williams, "Investigation," 14.

102 Symonds and Williams, "Probability of Return to Full Flying," 1.

103 Privy Council Office, "Report," 65.

104 Symonds and Williams, "Clinical and Statistical Study," 33; Noy, "Combat Stress Reactions," 508.

105 Copp and McAndrew, *Battle Exhaustion*, 58, 81, 114. Copp and McAndrew, and the original compilers of the data, readily acknowledged that the army statistics contained errors attributable to the difficulties of data collection under battlefield conditions. See, for example, 57–8, 80, 217–18 n.29.

106 English, *On Infantry*, 138. Infantry rates are cited here because they were the highest in the British and American armies (three to ten times those of other combat arms).

107 This is a rough estimate based on the following information: first, that the average aircrew strength of a Group in 1942 was approximately 1,160; second, that there were seven operational Groups in February 1943: Rexford-Welch, ed., *Royal Air Force Medical Services*, 2: 123; Webster and Frankland, *Strategic Air Offensive against Germany*, 4: 403–6. A typical nighttime bombing operation could involve 4,000 to 5,000 aircrew: Verrier, *The Bomber Offensive*, 205.

108 Harvey, *Boys, Bombs and Brussels Sprouts*, 159.

109 Peden, *A Thousand Shall Fall*, 416; Longmate, *The Bombers*, 188; Schurman interview with the author.

110 Hull interview with the author. General Hull commanded Bomber Command's 428 Squadron. When his flight engineer asked to be removed from flying duties because of worries about his sick mother, he was removed but labelled LMF. Hull thought the LMF procedure was fairer than a court martial.

111 Lawson quoted in Terraine, *The Right of the Line*, 532.

CHAPTER SIX

1 McCarthy, "Aircrew," 87; Rexford-Welch, ed., *Royal Air Force Medical Services*, 2: 128.

2 Ward, ed., *A Party Politician*, 202, 204.

3 Dunmore and Carter, *Reap the Whirlwind*, 254–5; Harvey, *Boys, Bombs and Brussels Sprouts*, 163.

4 DHIST, 181.009 (D614), RCAF Overseas Headquarters file S.2-11-2, vol. 1, minute 3, Leaman to Deputy Director of Personnel 1 (DDP1), 12 July 1945; HQ 840-108, vol. 5, "Disposal and Treatment of Personnel Removed from Aircrew Duties," 18 November 1942. The HQ 840-108 files are the property of the National Archives of Canada, but are currently on long-term loan to DHIST.

5 PRO, AIR 33/5, E.R. Ludlow-Hewitt, Inspector General's Report No. 271, 2; Appendix I, 1; Attachment B.1 to List A.1, 1; List A.2, 3. See chapter 5 for a more complete account of this practice.

6 DHIST, 181.003 (D132), "Report, Minister of National Defence for Air, Mission to United Kingdom, 30-6-41 to 24-7-41" [hereafter: MNDA Report], 1.

7 Suspended NCO aircrew who had a trade before remustering to aircrew duties were allowed to resume their previous trade and rank if they were found to be LMF: "Notes of a general discussion held in AMP's room on Sunday, July 6th, 1941," 2, attached to DHIST, MNDA Report.

8 Draft letter from Air Ministry (unsigned) to Stevenson, 5 July 1941, attached to DHIST, MNDA Report. Air Commodore L.F. Stevenson commanded the RCAF Overseas until January 1942: Douglas, *Creation of a National Air Force*, 323.

9 DHIST, MNDA Report, 8.

10 PRO, AIR 2/8591, "Canada, Regulations Respecting Officers and Airmen Removed from Aircrew Duties," 23 October 1941, 1–2. There are two copies of these regulations in the PRO file, one at f. 165(c), and the other at f. 176A. The latter copy adds "Deprivation of Flying Badges, Regulations" to the title, and includes the text of the relevant amendment to the King's Regulations for the RCAF. Both copies were sent to the AOC RCAF Overseas on 3 November 1941. These regulations appear to have been received by the RAF's AMP around the middle of February 1942, although a copy was sent to him in December 1941: see PRO, AIR 2/8591, f. 165 (D), MacLean to AOC RCAF Overseas, 3 November 1941; f. 165 (b), minute, Jackson to AMP [RAF], 11 December 1941; f. 178B, loose minute, Principal Secretary (PS) to AMP, 17 February 1942.

11 DHIST, "Disposal and Treatment of Personnel."

12 DHIST, 181.003 (D3456), Parks and Vlastos to AMP [RCAF], "Morale Survey," (December 1942), memorandum 5, 1.

13 Ward, ed., *A Party Politician*, 193, 230–1.

14 Ward, ed., *A Party Politician*, 24, 42–8. Ralston, too, had served overseas in World War I "with a very distinguished record": 193.

15 Power did not like the term Canadianization, for it implied an unyielding policy of making things Canadian, whereas he saw Canadianization as more of a gradual, negotiated change, the circumstances of the war permitting: Ward, ed., *A Party Politician*, 214.

16 Carter, *RAF Bomber Command and No. 6 [Canadian] Group*, 129–30; Douglas, *Creation of a National Air Force*, 249–50; Ward, ed., *A Party Politician*, 229.

17 Queen's University Archives (QUA), Charles Gavan Power Papers, box 64, D-1084, John McNab, "Interim Report – Visit Principal Chaplain (P)," 5 July 1942.

18 Ward, ed., *A Party Politician*, 220–1.

19 The officer in charge of RCAF Overseas Headquarters had a number

of titles, including AOC and Air Officer-in-Chief (AO-in-C) RCAF Over-
seas. Later this was changed to Air Officer Commanding-in-Chief
(AOC-in-C) RCAF Overseas. I have used AOC-in-C throughout the text.
For more information see Stacey, *Arms, Men and Governments*, 531.

20 Haycock, *Hughes*, 313.
21 QUA, Power Papers, box 64, D-1084, Edwards to de Carteret, 6 January
   1942, 3.
22 PRO, AIR 2/8591, f. 165(b), Jackson to AMP [RAF], 11 December 1941.
   Emphasis in original. At a meeting at RCAF Overseas Headquarters
   held on 12 December 1941 (note 23 below), it was also remarked that
   policy did not require disposal of disciplinary cases by RCAF authorities.
23 DHIST, 181.003 (1172), "Comments of [sic] a meeting held in Group
   Captain Curtis' Office RCAF Overseas HQ, on 12th December, 1941," 1.
24 DHIST, "Disposal and Treatment of Personnel."
25 PRO, AIR 2/8591, f. 167(a), AMP [RAF] to AO-in-C RCAF Overseas, 15
   December 1941.
26 In both the RAF and the RCAF the senior policy-making body was
   known as the Air Council. At first the RCAF version comprised the
   divisional heads at AFHQ: Douglas, *Creation of a National Air Force*,
   138. Later it more closely imitated the RAF model, and included
   Cabinet ministers with air force responsibilities, senior civil servants,
   and the most senior air officers at AFHQ.
27 PRO, AIR 2/8591, f. 167 (a), "Notes of a meeting held on 30th December, 1941, to discuss the conditions under which aircrew distinguishing badges are withdrawn from personnel classified lacking in moral
   fibre and arrangements for the subsequent disposal of these personnel," 1–3.
28 PRO, AIR 2/8591, f. 172A, "Notes of a meeting held on 5th February,
   1942, to consider AMP's objections to a recommendation for the
   amendment of the Air Council regulations governing the withdrawal
   of distinguishing badges," 1–3. Emphasis in original.
29 PRO, AIR 2/8591, f. 185A, Evans to All Commands and Groups, Home
   and Overseas, 4 March 1942; f. 188A, Evans to AOC-in-C RCAF Overseas, 9 March 1942.
30 DHIST, 181.003 (D4761), "Minutes of RCAF Squadron Adjutants' Conference Held at RCAF Overseas HQ on Friday, May 29th 1942," 11–12.
31 QUA, Power Papers, box 64, D-1084, Edwards to Breadner, Report No.
   1, 21 February 1942, 6. At first Power "refused" to give Edwards the
   authority to approve the removal of flying badges and the demotion
   of NCO aircrew, but Power appears to have relented after an appeal
   from Edwards: see Edwards to Breadner, Report No. 11, 12 January
   1943, 1.

32 It was also agreed that once the Canadian procedures were fully in place, "the actual removal of the badge or reduction in rank [would] be carried out by a Canadian Officer": DHIST, 181.003 (D4761), "Minutes of Squadron Commanding Officers' Conference Held at RCAF Overseas HQ on Friday, March 6th, 1942," 16–17; DHIST, "Minutes of RCAF Squadron Adjutants' Conference," 12.

33 DHIST, "Minutes of RCAF Squadron Adjutants' Conference," 11–12.

34 DHIST, "Minutes of RCAF Squadron Adjutants' Conference," 11; "Minutes of Squadron Commanding Officers' Conference," 17.

35 PRO, AIR 2/8592, minute 3, AMP to P.9 and S.7, 20 April 1942; minute 4, Director of Personal Services (DPS) to AMP, 1 May 1942; minute 5, AMP to DPS, 1 May 1942; minute 6, AMP to DPS, 1 May 1942; minute 26, comments of S.7, 1 September 1942; minute 35, comments of S.7, 22 October 1942.

36 Ward, ed., *A Party Politician*, 233.

37 Balfour, *Wings over Westminster*, 161.

38 Balfour's account suggested that he began reviewing LMF cases sometime in 1942: *Wings over Westminster*, 193–6. However, the minutes of a meeting held on 20 October 1944 quote him as saying that "[h]is only contact with such cases was in connection with enquiries addressed to him by members of the House." The minutes further record that it was only at this meeting that Balfour was first asked to review all LMF cases involving airmen: PRO, AIR 2/8592, "Minutes of a meeting held on the 20th October 1944, to discuss the 'w' procedure," 3–4.

39 Balfour, *Wings over Westminster*, 50, 161, 195–6.

40 For a discussion of the Ottawa Air Training Conference see Douglas, *Creation of a National Air Force*, 250–60 (quotation at 249–50).

41 Douglas, *Creation of a National Air Force*, 260.

42 DHIST, HQ 840-108, vol. 4, "Minutes of Special Meeting of Air Council and RCAF Delegates to the Air Training Conference ... ," 22 May 1942, 2; DHIST, "Minutes of a Squadron Commanding Officers' Conference," 16.

43 QUA, Power Papers, box 59, D-1041, "New Draft with Amendment, Deprivation of Flying Badges for Disciplinary and Other Reasons" (29 May 1942), 1, attached to Memorandum, Middleton to Minister, 30 May 1942. Emphasis in original.

44 QUA, Power Papers, box 59, D-1041, Copy of Air Council Regulation S.61141.

45 Terraine, *The Right of the Line*, 533.

46 QUA, Power Papers, box 59, D-1041, Note to Undersecretary of State for Air [US of S] (c) from H[ollinghurst], "Deprivation of Flying Badge," 1 June 1942, 2.

47 QUA, Power Papers, Memorandum, Middleton to Minister, 30 May 1942; DHIST, HQ 840-108, vol. 5, Minutes of Air Council Meeting No. 23, 3 December 1942, 2–3 and Appendix "A."

48 Douglas, *Creation of a National Air Force*, 280.

49 PRO, AIR 2/8592, minute 10, S.7 to Handley, 3 June 1942; minute 13, AMP to DPS and S.7, 23 June 1942; minute 16, DPS to AMP, 9 July 1942. Perhaps Babington's lack of understanding of the Dominion viewpoint was due to the fact that, according to Power's records, most of these matters were discussed with Hollinghurst in Babington's absence: see QUA, Power Papers, box 59, D-1041.

50 PRO, AIR 2/8592, "Notes of a meeting held in AMP's room on 23rd July, 1942, to discuss the regulations governing the withdrawal of the flying badge and certain aspects of the 'w' procedure."

51 PRO, AIR 2/8592, Matthews (for AOC-in-C RCAF Overseas) to S.7(c), 4 August 1942.

52 QUA, Power Papers, box 64, D-1084, Edwards to Breadner, Report No. 8, 3 October 1942, 7.

53 PRO, AIR 2/8592, McNamara to US of S, 17 August 1942; minute 25, S.7C.1 to AMP, 31 August 1942.

54 PRO, AIR 2/8591, McNamara to PS to AMP, 14 November 1941; minute 27, comments of S.10, 4 September 1942; minute 31, DPS to AMP, 10 September 1942 (quotation from minute 27). By May 1942 the RAF had recognized the manpower problem and allowed medically unfit aircrew (except LMF cases) to remuster to ground trades that required aircrew experience: DHIST, 181.009 (D3354), A.339936/41/S.10(b), R.B. Richards (for US of S) to ?, 16 May 1942 (copy).

55 PRO, AIR 2/8592, S.61141/S.10(b), Air Ministry to Distribution [all RAF and Dominion units], 25 February 1943.

56 QUA, Power Papers, box 64, D-1084, Edwards to Breadner, Report No. 1, 1; Report No. 2, 30 March 1942, 5; Stacey, *Arms, Men and Governments*, 272.

57 QUA, Power Papers, Edwards to de Carteret, 6 January 1942, 2; Edwards to Breadner, Report No. 1, 1.

58 QUA, Power Papers, Edwards to Breadner, Report No. 8, 7.

59 DHIST, "Disposal and Treatment of Personnel."

60 Ward, ed., *A Party Politician*, 5; QUA, Memorandum, Middleton to Minister, 30 May 1942.

61 DHIST, Minutes of Air Council Meeting No. 23, 2–3 and Appendix "A."

62 DHIST, HQ 840–108, vol. 5, Minutes of Air Council Meeting No. 24, 10 December 1942, 4–5 and Appendix "A."

63 DHIST, "Disposal and Treatment of Personnel."

64 DHIST, Minutes of Air Council Meeting No. 24, Appendix "A."

65 QUA, Power Papers, box 64, D-1084, Curtis to Breadner, Report No. 10, 1 December 1942, 1; Edwards to Breadner, Report No. 11, 1.

66 PRO, AIR 2/8592, Wait (for AOC-in-C RCAF Overseas) to P.3 (Can.) [Air Ministry], 5 March 1943. This letter was signed by Squadron Leader Gordon Screaton.

67 PRO, AIR 2/8592, Monk Jones to AOC-in-C RCAF Overseas, 31 March 1943; Wait (for AOC-in-C RCAF Overseas) to US of S (attention S.7.C.1), 7 April 1943.

68 They were Squadron Leaders Gordon Screaton and P.H. Cunningham. Screaton was "[l]iaison with Air Ministry on casualties, service estates, etc.": DHIST, "Minutes of RCAF Squadron Adjutants' Conference," Appendix A, 3.

69 PRO, AIR 2/4935, "Notes of a Meeting held at Air Ministry on Wednesday, 17th March, 1943, to consider a revised edition of Air Ministry Memorandum S.61141/S.7.C(1) dated 19th September, 1941"; "Memorandum on the Disposal of Members of Air Crews Who Forfeit the Confidence of their Commanding Officers," (1 June 1943).

70 Lawson, for example, stated that he believed the RCAF had agreed to "our policy for the removal of badges at the OTU stage for any reason," when it had not. To test the waters, Lawson "put over a test case to them on the 23rd March": PRO, AIR 2/8592, minute 2, Lawson to AMP, 18 April 1942. See also McNamara to US of S, 17 August 1942; McNamara to PS to AMP, 14 November 1941.

71 See the correspondence in PRO, AIR 2/4935, 8591–2.

72 Ward, ed., A Party Politician, 206.

73 DHIST, "Minutes of Special Meeting of Air Council and RCAF Delegates," 2.

74 PRO, "Minutes of a meeting held on the 20th October 1944," 1. Neither Balfour nor Sinclair appears to have participated in the discussions to the extent that Power did.

75 DHIST, 181.009 (D616), Massey to Acting Air Officer Commanding-in-Chief (A/AOC-in-C), "Tour of Duty at RCAF 'R' Depot, Warrington ... ," 14 November 1943, 2.

76 PRO, AIR 2/4935, Consultants in Neuropsychiatry, "Comments on the Memorandum on the Disposal of Members of Aircrews who Forfeit the Confidence of their Commanding Officers, S.61141/S.7C(1) of 1st June 1943"; minute 60, Director-General of Personal Services to AMP, 1 March 1944.

77 PRO, AIR 2/8592, "Memorandum of [sic] the Disposal of Members of Aircrews who Forfeit the Confidence of their Commanding Officers, S.61141/S.7(d)," (1 March 1945), especially paragraph 15 (b).

78 PRO, "Minutes of a meeting held on the 20th October 1944," 4.

79 Air Ministry to Stevenson, attached to DHIST, MNDA Report.

80 Terraine, The Right of the Line, 534.

81 Hastings, Bomber Command, 244–5.

82 PRO, "Minutes of a meeting held on the 20th October 1944," 4.

83 Hastings, *Bomber Command*, 248–9.

84 Longmate, *The Bombers*, 188.

85 Fernandez-Armesto, "The Experiences of Officers," 11–12, 15–16, 18–20, 24, 29, 32; Mason, *History*, 7. Officers were advised that the "'great point in running a unit or station is to inculcate a public school feeling'": "Wing Commander" quoted in Fernandez-Armesto, "The Experiences of Officers," 24.

86 Sir John Slessor quoted in Ward, ed., *A Party Politician*, 234. A more complete discussion of this phenomenon is found in Ellis, *Social History of the Machine Gun*, 104–7. Unlike Portal, who had had the benefit of a public-school education, Harris had had minimal formal education. His greatest accomplishment at boarding school seems to have been that he learned how to handle animals: Saward, *"Bomber" Harris*, 4–5.

87 Fernandez-Armesto, "The Experiences of Officers," 24.

88 Before the war, RAF officers were expected to wear "undress Mess kit" every night for dinner, except on guest nights, when local dignitaries and other guests of high social standing were entertained in the Mess and "full Mess kit" was required. Junior officers were advised, among other things, to "Never remain in a public bar if other ranks are present or enter," and to "avoid travelling in the same [rail] compartment as other ranks": Stradling, *Customs of the Service*, 16, 50–1. Group Captain Stradling first published his little book of "helpful hints and advice to those newly commissioned" in March 1939. The quotations are from the 1947 edition.

89 Bowyer, *2 Group RAF*, 27.

90 Harvey, *The Tumbling Mirth*, 66–8.

91 Bowyer, *2 Group RAF*, 28–9, 31.

92 Symonds and Williams, "Psychological Disorders ... Section 2," 10, 15.

93 Hastings, *Bomber Command*, 243.

94 Terraine, *The Right of the Line*, 534.

95 Dunmore and Carter, *Reap the Whirlwind*, 153–5.

96 "Note of a Meeting Held at the Air Ministry, London, on 8th July 1941 ... ," 3, in DHIST, MNDA Report.

97 Dunmore and Carter, *Reap the Whirlwind*, 154.

98 McKay, *One of the Many*, 92.

99 Ward, ed., *A Party Politician*, 224.

100 Boyle, *Trenchard*, 199; Lewis, *Aircrew*, 29.

101 Terraine, *The Right of the Line*, 464–5.

102 For a more complete discussion of the commissioning issue see Carter, *RAF Bomber Command and No. 6 [Canadian] Group*, 139–46.

103 QUA, Power Papers, Edwards to de Carteret, 6 January 1942, 2–3. Emphasis in original.

104 PRO, AIR 2/8591, MacNeece Foster to US of S, 7 March 1942.

105 DHIST, HQ 840–108, vol. 6, "Minutes of Meeting of Air Members of Air Council ... ," 15 April 1943, 2, and "Supporting Data for Item No. 8."

106 Douglas, *Creation of a National Air Force*, 270, 274.

107 Peden, *A Thousand Shall Fall*, 112.

108 Hastings, *Bomber Command*, 242–3.

109 Carter, *RAF Bomber Command and No. 6 [Canadian] Group*, xiii, 135.

110 Carter, *RAF Bomber Command and No. 6 [Canadian] Group*, 41, 91, 142, 155–61 (quotation at 41).

111 DHIST, "Comments of [sic] a meeting held in Group Captain Curtis' Office," 2; "Minutes of Squadron Commanding Officers' Conference," 11; "Minutes of RCAF Squadron Adjutants' Conference," 12.

112 When Edwards arrived in Britain, RCAF Overseas Headquarters had no suitable records system to keep track of the personnel under its authority. One was not established until the summer of 1943: QUA, Power Papers, box 64, D-1084, Edwards to Breadner, Report No. 1, 2; Report No. 15, 19 July 1943, 6.

113 DHIST, 181.003 (D1172), Message, Edwards to Breadner, 16 February 1942.

114 PRO, Inspector General's Report No. 271, 1, 4; DHIST, 181.003 (D1172), Headquarters Flying Training Command to [Air Ministry, all Commands, and RCAF Overseas Headquarters], 23 July 1942; 181.009 (D613), Burden to Director of Personnel (DP), RCAF Overseas Headquarters, 5 September 1942.

115 RCAF Warrington was referred to as a personnel disposal centre and as an "R" or repatriation depot. It was located about eighteen miles east of Liverpool. In May 1945 the depot was moved to Torquay, Devonshire: DHIST, 181.009 (D616), Air Ministry 683/A.D.O.1., Air Ministry to AOC-in-C Flying Training Command and others, 2 May 1945.

116 DHIST, 181.003 (D1172), Director of Air Staff to DP, Director of Accounts, and Principal Medical Officer, 24 August 1942; Edwards to Secretary Department of National Defence for Air, 29 September 1942; DHIST, 181.009 (D614), DP to Deputy Air Officer Commanding-in-Chief (DAOC-in-C), 10 September 1942; minute 3, P.1 to DDP1, 12 July 1945. With the opening of a Canadian medical board in London on 1 May 1944, medical boards were no longer conducted at Warrington: DHIST, 181.009 (D616), Noble (for AOC-in-C RCAF Overseas) to CO RCAF "R" Depot, 2 May 1944; Feasby, ed., *Official History* 2: 366–9.

117 DIST, 181.009 (D613), RCAF Overseas Headquarters file 2-1-1, "Memorandum" (unaddressed and unsigned), 18 September 1942.

118 Denton Massey was the brother of the Canadian High Commissioner in London, Vincent Massey. Denton Massey was sent to Britain to command the "R" Depot. During his time overseas he also was assigned other duties and prepared a number of special reports for Edwards: DHIST, Massey to A/AOC-in-C, 14 November 1943, 8; Stacey, *Arms, Men and Governments*, 299. A third brother, Hart, served as a flight lieutenant in RCAF Overseas Headquarters: Harvey, *Boys, Bombs and Brussels Sprouts*, 205.

119 DHIST, 181.009 (D616), Massey to AOC-in-C RCAF Overseas, 14 February 1943, 1, 8, Appendix C. The problem of poor record-keeping and record-transmission was to plague this system for several years: PRO, Inspector General's Report No. 271, 3, 6.

120 Edwards was alert to the fact that effective redeployment of manpower in the United Kingdom, when possible, avoided the long delays in receiving replacements from Canada: QUA, Power Papers, box 59, D-1034, "Minutes of Meeting No. 1 [between Power and senior officers, including the CAS and Edwards]," 26 May 1943, item 3.

121 DHIST, 181.009 (D613), Wait to US of S (attention Deputy Director of Personal Services 2), 22 April 1943; Venn to Wait, 25 May 1943; Wait to DAOC-in-C, 2 June 1943.

122 DHIST, 181.009 (D613), Wait to P.5, 15 September 1943; Venn to Wait, 16 September 1943; Wait to Venn, 30 September 1943.

123 QUA, Power Papers, box 64, D-1083, "[Minutes of a] Meeting held in the Minister's Office 23 February 1943 having under discussion a number of Overseas Problems," 3. Even though the reselection committee had operated since October 1943, an order establishing the committee was not signed by the AOC-in-C RCAF Overseas until 27 June 1944: DHIST, 181.009 (D613), RCAF Overseas Headquarters file 2-1-1, "Aircrew Re-selection Committee," 27 June 1944. The order establishing the Special Cases Committee was signed in March 1944: DHIST, 181.009 (D613), "[Order Appointing] Special Cases Committee," 9 March 1944; minute 3, P.1 to DDP1, 12 July 1945.

124 QUA, Power Papers, box 59, D-1035, "Minutes of a Conference with the Minister," 19 April 1944, 3; minute 3, P.1 to DDP1, 12 July 1945.

125 DHIST, 181.009 (D613), Edwards to CO RCAF "R" Depot, 11 May 1943; QUA, Power Papers, Edwards to Breadner, Report No. 15, 2.

126 PRO, Inspector General's Report No. 271, 3.

127 QUA, Power Papers, Edwards to Breadner, Report No. 15, 2. I have not been able to find complete statistics for the number of aircrew the boards at Warrington were able to salvage. Of 224 reported cases, 41 (18 per cent) were recommended for reassignment in the United Kingdom: DHIST, 181.009 (D613), CO "R" Depot to AOC-in-C RCAF Overseas, 3 September 1943; Massey to AOC-in-C RCAF Overseas, 11

January 1944; 181.009 (D616), CO "R" Depot to AOC-in-C RCAF Overseas, 3 May 1944.

128 QUA, Power Papers, box 64, D-1084, Edwards to Breadner, Report No. 6, 11 August 1942, 3. Edwards's wish was that personnel would "not stay loner than two weeks."

129 DHIST, 181.009 (D614), President Special Cases Committee to CO "R" Depot, 2 October 1945.

130 DHIST, Massey to AOC-in-C RCAF Overseas, 14 February 1943, 6.

131 DHIST, Massey to A/AOC-in-C, 14 November 1943, Appendix "B."

132 DHIST, 181.009 (D3354), Headquarters No. 6 (RCAF) Group to [all units in No. 6 Group], 12 April 1944; No. 62 (RCAF) Base to RCAF Stations Linton, Tholthorpe, and Eastmoore, and to Nos. 408, 415, 420, 425, 426, and 432 (RCAF) Squadrons, 20 November 1944.

133 DHIST, "Minutes of RCAF Squadron Adjutants' Conference," 8–9; DHIST, 181.009 (D3354), Headquarters No. 6 (RCAF) Group to [RCAF bases and stations in No. 6 Group], 22 September 1943.

134 DHIST, 181.009 (D3354), Headquarters No. 6 (RCAF) Group to RCAF Stations Eastmoor, Leeming, Middleton St. George, and Topcliffe, 17 April 1943.

135 DHIST, 181.009 (D3354), Headquarters No. 6 (RCAF) Group to [all units in No. 6 Group] 12 May 1943. An instance was cited where a psychological case was sent to Warrington under the wrong authority and without the proper paperwork. The circular of 12 May 1943 reminded RCAF units that proper procedures had to be followed, and that "[f]ull executive reports" were required. The circular forms part of a file that contains administrative instructions clarifying, cancelling, and redefining orders and procedures.

136 DHIST, 181.009 (D616), AOC-in-C RCAF Overseas Headquarters to CO "R" Depot, 6 April 1944.

137 Longmate, *The Bombers*, 174.

138 Dunmore and Carter, *Reap the Whirlwind*, 61. PRO, AIR 8/733, Harris to CAS, 19 December 1942; AMT to CAS, 21 December 1942; Portal to Air Member for Supply and Organization, 24 December 1942.

139 DHIST, Parks and Vlastos "Morale Survey," memorandum 5, 2.

140 Carter, *RAF Bomber Command and No. 6 [Canadian] Group*, 58–9, 65–8; Dunmore and Carter, *Reap the Whirlwind*, 107, 143–4, 158, 176–7, 195–9, 297.

141 QUA, Power Papers, box 64, D-1090, Edwards to Breadner, Report No. 17, 25 September 1943, "Waverer Cases Reported by Five Heavy Bomber Groups in B[omber] C[ommand] in the Past Five [?] Months" (attachment to report).

142 The RCAF statistics presented here are from DHIST, President Special Cases Committee to CO "R" Depot, 2 October 1945. The RAF statistics

are from Symonds and Williams, "Occurrence of Neurosis ... 1944 and 1945," 1. The casualty figures are from Webster and Frankland, *Strategic Air Offensive against Germany*, 4: 440–1.

143 Balfour, *Wings over Westminster*, 195.

144 "Note of a Meeting ... 8th July 1941," 3, in DHIST, MNDA Report.

145 Ward, ed., *A Party Politician*, 5.

146 The Air Council drew attention to this fact when drafting the original RCAF regulations: DHIST, Minutes of Air Council Meeting No. 23, 2. Power emphasized that he did not want "our attitude" to be "unduly harsh," and that maximum use should be made of airmen unfit for full overseas duty: HQ 840-108, vol. 6, Minutes of Air Council Meeting No. 9, 4 March 1943, 4.

CHAPTER SEVEN

1 Stacey, *Arms, Men and Governments*, 52, 397. The political story has only been summarized in this chapter. The reader who requires more detail may wish to consult the works cited in the notes.

2 Arthur Marwick, "Problems and Consequences of Organizing Society for Total War," in Dreisziger, ed., *Mobilization*, 14; Burns, *Manpower*, 3.

3 Burns, *Manpower*, 130–1.

4 Granatstein and Hitsman, *Broken Promises*, 69.

5 Granatstein and Hitsman, *Broken Promises*, 98.

6 Quoted in Granatstein and Hitsman, *Broken Promises*, 98.

7 Granatstein and Hitsman, *Broken Promises*, 63, 69, 84, 98–9. Skelton said that the Military Service Act "'did more to win the election than to win the war'": quoted in Walker, "Poles Apart," 47.

8 The complete story is in Haycock, *Hughes*, especially 198–224.

9 Granatstein and Hitsman, *Broken Promises*, 95–6. The quotation is from Haycock, *Hughes*, 198.

10 Wise, *Canadian Airmen*, 87. RFC Canada experienced its own manpower crisis, but it arose from a dearth of qualified tradesmen: Wise, *Canadian Airmen*, 108–10.

11 Wise, *Canadian Airmen*, 634.

12 RG 24, vol. 2041, HQ 6978-2-131, vol. 2, f. 198, Gwatkin (CGS) to Hoare, 29 October 1917.

13 Winter, *The First of the Few*, 17. Winter has attacked the view that air forces were not important to the outcome of World War I. He has pointed out that they altered the whole face of battle by forcing armies to move at night to avoid observation: 12. Kennett has described the air forces' contributions as "immense" because they directed the "supreme killing device in the Great War ... the exploding artillery shell": Kennett, *The First Air War*, 220.

14 Stacey, *Arms, Men and Governments*, 71.

15 Stacey, *Arms, Men and Governments*, 43.
16 Burns, *Manpower*, 116.
17 Plumptre, *Mobilizing Canada's Resources*, 30.
18 Granatstein, *Canada's War*, 101.
19 Burns, *Manpower*, 116.
20 Morton, *Canada and War*, 115, 126.
21 Stacey, *Arms, Men and Governments*, 403. Owing to its relatively small size and unique attractions, the Royal Canadian Navy "never had much difficulty in recruitment": Granatstein and Morton, *A Nation Forged in Fire*, 69.
22 Granatstein, *Canada's War*, 9, 341 (quotation at 9, from a memorandum to Cabinet from the Chiefs of Staff, 29 August 1939). Stephen Harris has claimed, according to Walker, that Mackenzie King "knew of, and approved, the general staff's plan for the expeditionary force while intentionally keeping his Cabinet colleagues in the dark": Walker, "Poles Apart," 85.
23 Stacey, *Arms, Men and Governments*, 41–2.
24 Stacey, *Arms, Men and Governments*, 46–8.
25 Morton, *Military History of Canada*, 145.
26 Stacey, *Arms, Men and Governments*, 51–2.
27 Stacey, *Arms, Men and Governments*, 481–2.
28 Stacey, *Arms, Men and Governments*, 481–2.
29 Burns, *Manpower*, 125.
30 Granatstein, *Canada's War*, 373.
31 Walker, "Poles Apart," 113, 129.
32 Walker, "Poles Apart," 125.
33 Stacey, *A Very Double Life*, 200–4. See also Stacey, *Arms, Men and Governments*, 456–60.
34 Stacey, *Arms, Men and Governments*, 469–70, 472.
35 Stacey, *Arms, Men and Governments*, 201; Granatstein, *Canada's War*, 249. The other fundamental error identified by Stacey was Canada's dispersal of its overseas air forces: Stacey, *Arms, Men and Governments*, 201. Burns has disagreed that the two-theatre disposition caused undue administrative duplication: Burns, *Manpower*, 40–2.
36 Granatstein, *Canada's War*, 335, quoting Richard Hanson, the leader of the Conservative party in the House of Commons.
37 Burns, *Manpower*, 115; Stacey, *Arms, Men and Governments*, 44; Morton, *Canada and War*, 116–17, 140; Granatstein, *Canada's War*, 366–7; Walker, "Poles Apart," 47, 51, 81.
38 Stacey, *Six Years of War*, 1: 117.
39 English, *The Canadian Army*, 230.
40 English, *On Infantry*, 121, 138–9; Burns, *Manpower*, 6, 14, 23; Granatstein, *Canada's War*, 341.
41 Granatstein, *Canada's War*, 349.

42 Stacey, *Arms, Men and Governments*, 481.

43 Burns, *Manpower*, 85, 87, 93–4 (quotations at 87).

44 Morton, *Military History of Canada*, 219; Morton, *Canada and War*, 145.

45 Stacey, *Arms, Men and Governments*, 418.

46 Hitsman, "Manpower Problems," 24.

47 Hitsman, "Manpower Problems," 18.

48 Stacey, *Arms, Men and Governments*, 301, 417. The figure of 80,000 was given by the minister of labour in August 1943: DHIST, "RCAF Personnel History," 2: 375.

49 Hitsman, "Manpower Problems," 18, 20.

50 Plumptre, *Mobilizing Canada's Resources*, 13, 45.

51 Morton, *Military History of Canada*, 185; Plumptre, *Mobilizing Canada's Resources*, xvii.

52 Report of the Labour Supply Investigation Committee quoted in Stacey, *Arms, Men and Governments*, 403.

53 Terraine, *The Right of the Line*, 532 (citing J. Lawson); Morton, *Military History of Canada*, 187. Lawson stated "'that the pay of aircrew was very attractive indeed ... [and] has probably played a much greater part than is thought.'"

54 Robert Bothwell, "'Who's Paying for Anything these Days?' War Production in Canada 1939–45," in Dreisziger, ed., *Mobilization*, 68.

55 Stacey, *Six Years of War*, 1: 117.

56 In 1944 General Kenneth Stuart, the Chief of Staff of Canadian Military Headquarters in London, England (and a former CGS), "suppressed early warnings [of manpower shortages] and hoped for the best": Morton, *Canada and War*, 145.

57 Stacey, *Arms, Men and Governments*, 440; Burns, *Manpower*, 79. Burns has argued that further manpower wastage occurred in the army because "officers and men in the reinforcement stream were, generally speaking, junior and inexperienced, and not capable of training their men in ordinary subjects": 79.

58 Air Ministry, *Flying Training*, 1: 151–2, 223.

59 Air Ministry, *Flying Training*, 1: 243.

60 Douglas, *Creation of a National Air Force*, 292.

61 Air Ministry, *Flying Training*, 1: 269.

62 Burns, *Manpower*, 7, 166.

63 Hitsman, "Manpower Problems," 46.

64 Bezeau, "DHist Interim Report," 3.

65 Bezeau, "DHist Interim Report," 3, 5, 6.

66 Feasby, ed., *Official History*, 2: 456.

67 An "approach" was the term the recruiters used to describe a contact made by someone expressing an interest in joining the service.

68 DHIST, "RCAF Personnel History," 2: 415. The figure of 6.5 per cent is based on recruiting statistics from 1 October 1939 to 30 June 1944.

69 Douglas, *Creation of a National Air Force*, 293.
70 DHIST, "History of the AMP Division," 104.
71 Bezeau, "DHist Interim Report," 8.
72 The figure of 60 per cent is from Roberts, *There Shall Be Wings*, 236. In 1942 AFHQ estimated that about "two-thirds of RCAF aircrew ultimately will be serving with RAF squadrons": DHIST, HQ 840-108, vol. 5, "Minutes of Meeting of Air Members of Air Council ... ," 27 July 1942, 2.
73 Bezeau, "DHist Interim Report," 9b.
74 Reid, "Prognosis," 238.
75 The LMF and neurosis percentages are those mentioned in chapter 5. The casualty figures are from Stafford-Clark, "Morale," 23.
76 This estimate is based on a comparison of two statistics, the first being that 72,835 RCAF aircrew graduates were produced by the BCATP from October 1940 to March 1945, and the second that 589,149 "approaches" were recorded by RCAF recruiting centres from October 1939 to 30 June 1944: Hatch, *Aerodrome of Democracy*, 206; DHIST, "RCAF Personnel History," 2: 415.
77 Stacey, *Arms, Men and Governments*, 259 (quoting J.L. Ralston). Ralston's December 1940 estimate was borne out in May 1945, when aircrew represented 22 per cent of the RAF's total strength: Terraine, *The Right of the Line*, 681. At one heavy-bomber station there were 1,000 ground crew to 200 aircrew, according to Terraine: 467.
78 DHIST, "RCAF Personnel History," 2: 415.
79 Lightman, "Economics of Military Manpower Supply," 212. Even though Lightman's research was based on Canadian data from the 1960s, the same phenomenon was noted in Canada during World War II.
80 Morton, *Canada and War*, 137.
81 Morton, *Military History of Canada*, 179; Morton, *Canada and War*, 137. Less than 2 per cent of RCAF enrollees were unemployed when they enlisted: Bezeau, "DHist Interim Report," 5.
82 This certainly happened in World War I, when "thousands of Canadians" bypassed the infantry by joining other branches of the army: Morton, *Military History of Canada*, 136.
83 Kennett, *The First Air War*, 85.
84 Terraine, *The Right of the Line*, 578, 641–2. Terraine's criticism may be overly harsh, as the army was struggling, albeit with more-experienced staff officers, with many of the same training problems that bedeviled the air force.
85 English, *The Canadian Army*, 111.
86 Smith, *Code Word CANLOAN*, 1, 4, 25.
87 Burns, *Manpower*, 89.
88 Morton, *Military History of Canada*, 185.

89 Stacey, *Arms, Men and Governments*, 590; Morton, *Military History of Canada*, 185.

90 Terraine, *The Right of the Line*, 682.

91 Stacey, *Arms, Men and Governments*, 305; Morton, *Military History of Canada*, 207. The RCAF Overseas lost 12,266 men on flying operations and 1,906 in training accidents: Stacey, *Arms, Men and Governments*, 305. Fatal "battle casualties" in the Canadian Army from 1939 to 1945 totalled 17,683: Feasby, ed., *Official History*, 2: 426. Of the Canadians who served with the RCAF and the RAF outside of Canada, 17,101 never came home: Granatstein and Morton, *A Nation Forged in Fire*, 118.

CHAPTER EIGHT

1 McNeill, *The Pursuit of Power*, 339.

2 A fuller discussion of this issue can be found in McDonald, Johnston, and Fuller, eds., *Applications*, 9–43, and in Johnston, Fuller, and McDonald, eds., *Aviation Psychology*, 139–63.

3 See for example Loo, "Pilot Selection"; Spinner, "Using the Canadian Automated Pilot Selection System," i–v; D.R. Hunter, "Aviator Selection," in Wiskoff and Rampton, eds., *Military Personnel Management*, 129–67. Another way of estimating the value placed on a particular human resource is to see how much an organization is prepared to invest in training that resource. The cost of aircrew training is among the highest for all occupations in the Canadian Forces; in 1987 it cost approximately $660,000 to train a pilot to wings standard: Catano, "Canadian Automated Pilot Selection System," 13–15.

4 This comparison is based on the weight of an RCAF aviator of average height – my father – who weighed 140 pounds when he joined the RCAF as an aircrew candidate in 1943. His weight was equivalent to about 2,000 troy ounces, which multiplied by the price of gold in 1943 (US $35 per troy ounce) gives a result of US $70,000. Arthur Harris (cited in Terraine, *The Right of the Line*, 536) estimated that it cost £10,000 to train each member of a Bomber Command aircrew. In June 1943 the exchange rate for British to American currency was £1=US $4.03 (*New York Times*, 11 June 1943), which means that aircrew training cost US $40,300 per person. The cost of training a Canadian pilot in World War I was Cdn $9,835: Sullivan, *Aviation in Canada*, 57.

5 To offer one example, a Cambridge don who used sociological methods to good effect in overcoming high accident rates at an RAF fighter station was ordered "to stop work ... since 'Group psychology is the affair of the Group Medical Officer.'" He continued his work unoffi-

cially, with the support of successive station commanders, and apparently achieved excellent results: Paterson, *Morale*, 13, 122, 199, 202.

6 Van Creveld, *Technology and War*, 231.

7 Macphail's "central purpose" in this work was to rebut Herbert Bruce's accusations about inefficiency in the CAMC: Morton, *Military History of Canada*, 273.

8 Dunmore and Carter, *Reap the Whirlwind*, 331.

9 QUA, Power Papers, box 64, D-1084, Edwards to Breadner, Report No. 9, 8.

10 Stacey, *Arms, Men and Governments*, 287.

11 Murray, *Strategy for Defeat*, 223–4.

12 The reader will recall from chapter 3 that Harris was forced to withdraw "88 per cent of his pilots to feed OTUs and Conversion Units before completion of their first tour": PRO, AIR 2/4935, "The Operational Tour-Minutes of Meeting Held on Thursday, 4th February, 1943."

APPENDIX A

1 Based on a year commencing 10 February and ending the following 9 February.

2 Privy Council Office, "Report," 63–5, 95.

3 Symonds and Williams, "Clinical and Statistical Study," 39; Symonds and Williams, "Occurrence of Neurosis ... 1944 and 1945," 1.

4 Rexford-Welch, ed., *Royal Air Force Medical Services*, 2: 122.

5 Webster and Frankland, *Strategic Air Offensive against Germany*, 4: 440.

APPENDIX B

1 DHIST, 181.009 (D614), Squadron Leader N.J. Ogilvie, President Special Cases Committee to CO "R" Depot, Torquay, 2 October 1945, 1–2.

2 DHIST, 181.009 (D614), Group Captain J.A. Hutchison, CO "R" Depot to AOC-in-C RCAF Overseas, 2 October 1945.

APPENDIX C

1 Symonds and Williams, "Occurrence of Neurosis ... 1944 and 1945," 178. More complete data on psychological casualties is in appendix A.

2 Webster and Frankland, *Strategic Air Offensive against Germany*, 4: 441.

3 DHIST, 181.009 (D614), Squadron Leader N.J. Ogilvie, President Special Cases Committee to CO "R" Depot, Torquay, 2 October 1945, 2.

4 Webster and Frankland, *Strategic Air Offensive against Germany*, 4: 441.

Bibliographical Essay

In the search for sources, students of Commonwealth military aviation history are doubly fortunate to have available two excellent and indispensable works that summarize the literature of British military history in general and air force history in particular. Robin Higham's chapter on "The Development of the Royal Air Force, 1909–1945," included in his *Guide to the Sources of British Military History* (1971), provided a comprehensive survey of published materials up to 1971, and is still helpful to researchers interested in British and Canadian military aviation. In 1988 Stephen Harris and Norman Hilmer's update of Higham's essay appeared in Gerald Jordan's *British Military History: A Supplement to Robin Higham's Guide to the Sources*. The bibliographies accompanying the two essays identify, in total, over 600 items, and many of the bibliographical entries are annotated.

Although many books were written in the 1970s and 1980s on aspects of British air force history, Harris and Hilmer observed that no "comprehensive account of air force training or the way in which the RAF managed its personnel resources" had yet been published. They added that "questions about the composition of the air force, and personnel selection processes, are largely unanswered, while those relating to morale and discipline continue to be matters of conjecture." While Anthony Verrier's *The Bomber Offensive* (1974), Max Hastings's *Bomber Command* (1979), Norman Longmate's *The Bombers* (1983), and John Terraine's *The Right of the Line* (1985) had contributed to historians' understanding of manpower issues in the

RAF, Harris and Hilmer concluded that "[w]e still have a long way to go before our knowledge of the RAF as an institution and of its performance ... can in any way be considered complete."[1] Their comments remain valid today.

If RAF history can be said to be incomplete, then the history of the RCAF is by comparison still in its infancy. The number of scholarly works on the RCAF can be counted on the fingers of one hand, despite the immense influence the air force had on Canadian society and on Canada's contribution to the Allied war effort. The best works to date on the RCAF are the three volumes of the official air force history by S.F. Wise (volume 1), W.A.B. Douglas (volume 2), and Brereton Greenhous et al. (volume 3), which provide an excellent overview of the development of Canadian military aviation from 1914 to 1945 (the third volume appeared too late to be used in the preparation of this book). These works are notable for their clear prose, meticulous research, and exhaustive notes. However, because of the amount of ground they had to cover, the official historians did not address in detail the manpower issues raised by Hilmer and Harris. In this and other respects, RCAF history today is a generation behind the work done by historians of the Canadian Army.

William Carter was the first air force scholar to produce a well-researched and fully documented book examining some of the unanswered questions of RCAF history. His doctoral dissertation, published as *Anglo-Canadian Wartime Relations, 1939–1945, RAF Bomber Command and No. 6 [Canadian] Group* (1991), is an operational and social history of the RCAF's largest overseas formation in World War II. *Reap the Whirlwind* (1991), co-authored with Spencer Dunmore, adds personal anecdotes of those who participated in the bomber campaign to the account presented in Carter's dissertation, but otherwise offers no new material. These two works were the first to explore social aspects of the Canadian experience in Bomber Command, but like the official histories their broad scope precluded detailed discussion of personnel matters such as aircrew selection and the psychological aspects of combat.

The dearth of scholarly works on RCAF history stands in contrast to the many popular accounts of Canadian air force history that are available, such as Larry Milberry's *Aviation in Canada* (1979) and J.A. Foster's *For Love and Glory* (1989). Although they perform the valuable task of keeping the subject before the public, their contribution to historical knowledge is extremely limited. Published memoirs of the survivors of the air war of 1939–45 are becoming more numerous, but it is hard to quarrel with Harris and Hilmer's assessment that Murray Peden's *A Thousand Shall Fall* (1979) and Walter Thompson's

Lancaster to Berlin (1987) are two of the best personal accounts of the bomber campaign.[2] J. Douglas Harvey in *Boys, Bombs and Brussels Sprouts* (1981), *The Tumbling Mirth* (1983), and *Laughter-Silvered Wings* (1984) has performed yeoman service in gathering together short tales and anecdotes told by Canadian airmen, material that otherwise probably would not have seen the light of day. Merrily Weisbord and Merilyn Simonds Mohr, the authors of *The Valour and the Horror* (1991), the book published to accompany the controversial film of the same name, follow the film in using material on Canadians in Bomber Command that has appeared elsewhere, yet claim, in the words of two of their critics, "that they [are] in exclusive possession of an absolute truth, long hidden by some form of collective plot."[3] Although the producers have stated that the film and the book are based on detailed research,[4] the absence of citations in the book makes this assertion impossible to verify. Glaring errors, such as the claim that "Air Vice Marshall [sic] Harris sent the 'sonofabitch Edwards' [Air Vice-Marshal Harold Edwards, AOC-in-c RCAF Overseas] back to Canada,"[5] when, in fact, he did no such thing, call into question the accuracy of the research. Whatever one's views on *The Valour and the Horror* – a book that supposedly tells "The Untold Story of Canadians in the Second World War" – the story of Canadians in Bomber Command has certainly been told before, and in ways that provide more context and balance to events that occurred fifty years ago.

Even with the recent growth in interest in RCAF history, there remains a need for a critical historical analysis that measures air force operational performance against training and doctrine, along the lines of John English's *The Canadian Army and the Normandy Campaign* (1991). However, a number of works have looked at various aspects of training. F.J. Hatch's *Aerodrome of Democracy* (1983) discusses the history of the BCATP, but its length, 223 pages, limits it to a mere sketch of that enormous undertaking. Peter Conrad's *Training for Victory: The British Commonwealth Air Training Plan in the West* (1989) and Ted Barris's *Behind the Glory* (1992) provide an abundance of reminiscences, and look at some obscure, but fascinating, episodes (such as, in Barris, the making of the film *Captains of the Clouds*), but except for the personal anecdotes, the authors rely entirely on secondary sources for their information. More research on the history of aircrew training is required because of the subject's relevance to present-day concerns. In the name of economy, the Canadian Forces has just returned primary pilot training to private flying schools, as happened during the interwar years.[6]

While the history of RCAF training has been addressed to some extent, aircrew selection and psychological issues in air warfare are two areas that remain virtually unexplored. General works on personnel selection, most often written by medical doctors, scientists, or psychologists, have surveyed small parts of the territory from their own perspectives. Of these, the two most concerned with aviator selection, Douglas H. Robinson's *The Dangerous Sky: A History of Aviation Medicine* (1973) and T.M. Gibson and M.H. Harrison's *Into Thin Air: A History of Aviation Medicine in the RAF* (1984), recount, in broad terms, elements of the story of personnel selection in both world wars, glancing occasionally at their areas of special interest, aviation medicine in Robinson's case and aeromedical research in Gibson and Harrison's case. Some specialized studies of the psychological aspects of aircrew selection have appeared recently, most notably David Hunter's "Aviator Selection" in M.F. Wiskoff and G.M. Rampton's *Military Personnel Measurement* (1989) and Thomas Hilton and Daniel Dolgin's "Pilot Selection in the Military of the Free World" in Gal and Mangelsdorff's *Handbook of Military Psychology* (1991). In Canada little has been published in the field, but the Canadian Psychological Association is currently debating the legitimate role of psychology in the military.[7] Perhaps more published descriptions of the historical dimension of the application of psychology to practical military problems will provide a context for this discussion.

The complete story of personnel selection in the Canadian forces during the two world wars has yet to be written. Although two prominent Canadian psychologists were deeply involved in RAF personnel research, the British histories say little about their achievements. P.E. Vernon and J.B. Parry's *Personnel Selection in the British Forces* (1949), despite being over forty years old, is still a useful source for the history of many aspects of British aircrew selection, but, as with all of the works on this subject, it pays scant attention to important developments in Canada. H.S.M. Carver, a Canadian Army officer involved in personnel selection, produced "Personnel Selection in the Canadian Army," a DND internal study, in 1945. Nothing further of consequence appeared on personnel selection in the army until Terry Copp and Bill McAndrew took up some aspects of the subject in their book *Battle Exhaustion: Soldiers and Psychiatrists in the Canadian Army 1939–1945* (1990). Geoffrey Hayes examines selection in the Canadian Army officer corps in World War II,[8] but no study integrating research on the personalities, policies, procedures, and results of the selection process in the army has been undertaken. The history of Canadian aircrew selection has been even more poorly served, indeed largely ignored except for brief mentions in the official air force histories. There is thus still much research to be done, especially regarding the controversy over the use of the Link Trainer in pilot selection. Most of the original sources that relate to aircrew selection remain untouched.

As with aircrew selection, many questions about the history of human behaviour in air warfare have also been neglected. It may be that the larger field of aviation medicine "is finally receiving some attention,"[9] but the psychological and psychiatric aspects of Commonwealth air force combat have not received much attention. Although Gibson and Harrison's book has a chapter entitled "Coffins or Crackers" – the Bomber Command expression for being between the hammer of death and the anvil of neurosis – it has little to say about the mind itself, for it deals mainly with selection procedures and physiological problems (such as airsickness) as they relate to stress. The book is useful, however, for its references to World War II reports and other sources in the literature, despite several errors in the citations. Robinson's book, on the other hand, does look at "flying stress," but because it is a history of aviation medicine from the specialty's earliest days to the 1970s, its remarks on flying stress are confined to five pages and focus on the USAAF.[10] What the RCAF's experience was remains open to speculation.

Most accounts of the handling of World War II aircrew who would not or could not fly are, with the exception of Terraine's *The Right of the Line*, based on anecdotal material. And while Terraine's examination of the subject is the most carefully researched, it relies heavily on official RAF opinion, which colours his judgments. The present study seeks to look behind official opinion by drawing on a wider variety of primary sources than are found in Terraine.

Combat Stress Reaction, one of the main causes of psychological casualties in war, is still highly relevant to the Canadian military today, as has been demonstrated in recent peacekeeping operations. However, we do not have a sound historical perspective on the subject, nor have psychologists managed, it seems, to develop correct treatment protocols. Until the publication in 1980 of the third edition of the *Diagnostic and Statistical Manual of Mental Disorders* (DSM-III), the mental health professional's bible, which finally accepted posttraumatic stress disorder as a mental illness, veterans could experience the symptoms of but could not officially suffer from what was an unrecognized disease. Despite apparent advances in understanding CSR, the failed attempts at treatment by the Israelis in Lebanon, and by other nations during recent United Nations operations in the former Yugoslavia, indicate that there is still much to learn.[11] It may be that studies of the historical experience of CSR will give us some clues about which lines of inquiry might prove most fruitful.

In addition to the more recent works discussed so far, accounts written during or soon after the two world wars have also proved useful in preparing this book. One of the best works on RFC/RAF Canada is Alan Sullivan's semi-official history *Aviation in Canada 1917–18* (1919); among other things, it contains some superb photographs. The difficulty in obtaining this book is partly compensated for by the availability of William Chajkowsky's *Royal Flying Corps: Borden to Texas to Beamsville* (1979), an apparent reprint of Sullivan's book, published without credit to the original author.

Unpublished British Air Ministry records have been used extensively in various works, including the Air Historical Board narratives and the later official histories such as Charles Webster and Noble Frankland's chronicle of the RAF's bomber offensive against Germany. More recently, authors such as Terraine and Carter have carefully documented their use of Air Ministry sources. All of the works just mentioned have been extremely valuable in the preparation of the present book.

In Canada, recent scholarship has benefited enormously, as J.L. Granatstein has observed, from "access to all the records of Canada's role in the two world wars."[12] It is one thing, however, to have access, but quite another to be able to make good use of it, as John English has suggested by his reference to the DND files in RG 24 at the National Archives of Canada (NAC) as "either a gold mine or a quagmire."[13] Working in RG 24 can also be compared to panning for gold; one has to go through a lot of gravel to get the nuggets, and even great historians have had difficulty finding the treasure. This was pointed out to me by Barbara Wilson, a senior archivist at the NAC, who said that C.P. Stacey had, in some cases, to rely on unofficial accounts because of the dreadful state of the RCAF files.[14] The archival chaos has been partly rectified, but the bulk of the RG 24 air force material remains untapped, except by researchers at the DHIST.

In addition to what is kept at the NAC, the DHIST has its own collection of invaluable primary materials on Canadian air force history, easily accessible by scholars. No research on the RCAF's past can be complete without consulting the DHIST records.

Also valuable are the reports of the FPRC. Some have been published in the official Air Ministry study *Psychological Disorders in Flying Personnel of the Royal Air Force* (1947), and others in E.J. Dearnaley and P.B. Warr's *Aircrew Stress in Wartime Operations* (1979). While it is useful to have many of the most significant reports available without having to visit the archives, the Air Ministry work – the more extensive of the two collections – appeared shortly after the war, and its reprinted material was not accompanied by any synthesis or analysis of the FPRC's findings, nor was any attempt made to reconcile disagreements in the findings or to situate the data in the context of the 1939–45 experience. This was partly because most of the FPRC researchers left the RAF shortly after the war to pursue other interests. Dearnaley and Warr, similarly, included in their book only a short explanatory introduction to the FPRC reports by D.D. Reid, an author of some of the reports.

Lord Moran's book, *The Anatomy of Courage*, which by some is considered the classic work on courage in battle, was first published in 1945, and offered an interesting look at the opinions of senior medical people. But while he furnished colourful anecdotes to bolster his own theories, Moran ignored most of the scientific data collected by his medical colleagues. T.T. Paterson, a fellow of Trinity College, Cambridge, studied behaviour in combat from a sociological perspective, but his research was dismissed by RAF medical authorities. His *Morale in War and Work* (1955) shows what might have been accomplished had other paths of inquiry been followed in the quest to find solutions to psychological problems among aircrew.

Scholarly works on Canadian military history are generally of high quality and readily available. Stacey's seminal *Arms, Men and Governments* (1970) examines the history of manpower issues in the Canadian armed forces from military and political perspectives. Canada's conscription crises have been described in detail, and Desmond Morton's annotated bibliography in *A Military History of Canada* (1985) inventories the various works on the subject. To the works listed in Morton I would add R.J. Walker's unpublished MA thesis, "Poles Apart: Civil-Military Relations in the Pursuit of a Canadian National Army" (1991), which examines the army's motives in the national debate on obligatory military service that took place from 1867 to 1964, and provides evidence to show that many senior officers in the Canadian Army preferred some form of national service as a way of creating a national army.

Finally, for those readers who are wondering why the present study does not include more references to personal details from service records, the answer is – the Privacy Act. This legislation severely restricts access to the service's personnel records of the period examined here. In-depth analysis of their contents is a task for the next generation of historians of military aviation.

NOTES

1 Stephen Harris and Norman Hilmer, "The Development of the Royal Air Force, 1909–1945," in Jordan, ed., *British Military History*, 345.

2 Harris and Hilmer, "Development of the Royal Air Force," 353.

3 Bercuson and Wise, *The Valour and the Horror Revisited*, 9.

4 "CBC Ombudsman's Report was a Miscarriage of Justice: Producers," *Kingston Whig-Standard*, 25 November 1992.

5 Weisbord and Mohr, *The Valour and the Horror*, 78.

6 Kitchen, "Canada's Air Force," 3.

7 Wilson and Wenek, "A Clear Conscience." This recent contribution to the debate summarizes the major issues.

8 Hayes, "Science and the Magic Eye."

9 Harris and Hilmer, "Development of the Royal Air Force," 354.

10 Robinson, *The Dangerous Sky*, 182–7.

11 Copp, "From Flanders Fields to Sarajevo."

12 Quoted in Marc Milner, "Foreword," in Milner, ed., *Canadian Military History*, v.

13 English, *The Canadian Army*, 331.

14 Wilson interview with the author.

Bibliography

ARCHIVAL COLLECTIONS

NATIONAL ARCHIVES OF CANADA
Record Group 24 National Defence
Manuscript Group 28 Canadian Psychological Association

DIRECTORATE OF HISTORY, DEPARTMENT OF NATIONAL DEFENCE (OTTAWA)
Miscellaneous Papers, National Defence

NATIONAL RESEARCH COUNCIL OF CANADA
Reports

PUBLIC RECORD OFFICE (UNITED KINGDOM)
AIR 2 Air Ministry Registered Files
AIR 14 Bomber Command
AIR 20 Unregistered Papers (Air Ministry Headquarters Branches)
AIR 33 RAF Inspectorate General
AIR 41 Monographs and Narratives (RAF)
AIR 49 Medical Histories
AIR 57 DGMS – FPRC

QUEEN'S UNIVERSITY ARCHIVES (KINGSTON, ONTARIO)
Charles Gavan Power Papers

Norman Rogers Papers

UNPUBLISHED NARRATIVES

AIR HISTORICAL BOARD (UNITED KINGDOM)
"Flying Training – SFTSS 1934–42," n.d. (PRO, AIR 41/4).

DIRECTORATE OF HISTORY, DEPARTMENT OF
NATIONAL DEFENCE (OTTAWA)
"BCATP Flying Training," n.d. (DHIST 181.009 [D89A]).
Directorate of Manning, RCAF. 'Manning the RCAF,' (1944) (DHIST 74/15).
"Flying Training," n.d. (DHIST 74/19).
"The History of the AMP Division," 1944 (DHIST 74/14).
"RCAF Personnel History 1939–1945," n.d. 3 vols. (DHIST 74/7).

TRANSCRIPTS FROM THE ORAL HISTORY OF
PSYCHOLOGY IN CANADA
(National Archives of Canada, Manuscript Group 28)

Belyea, E., 1970
Chant, S.N.F., 1970
Myers, C.R., 1972
Signori, E.I., 1970
Springbett, B., 1970
Williams, D.C., 1969

INTERVIEWS WITH THE AUTHOR

Dr George Ferguson, Kingston, Ontario, June 1990
Chief Warrant Officer Raymond Gushue, Ottawa, June 1990
General Chester Hull, Ottawa, May 1992
Professor Donald M. Schurman, Kingston, Ontario, December 1992 and
    August 1993
Lieutenant-Colonel Karol W.J. Wenek, Kingston, Ontario, December
    1992
Ms Barbara Wilson, Ottawa, July 1990

OTHER SOURCES

Air Ministry (United Kingdom). *Flying Training*. Vol. 1, *Policy and Planning*
    (Air Publication 3233). London: HMSO, 1952.
– "Notes for Medical Officers on the Psychological Care of Flying Person-
    nel." 1st ed., May 1939 (PRO, AIR 2/8591).

– *Psychological Disorders in Flying Personnel of the RAF* (Air Publication 3139). London: HMSO, 1947.

Allred, Kenneth D., and Timothy W. Smith. "The Hardy Personality: Cognitive and Physiological Responses to Evaluative Threat." *Journal of Personality and Social Psychology* 56 (January 1989): 257–66.

Anderson, H. Graeme. *The Medical and Surgical Aspects of Aviation*. London: Henry Frowde, 1919.

– "The Selection of Candidates for the Air Service." *Lancet*, 16 March 1918, 395–9.

Armstrong, Harry G. *Principles and Practice of Aviation Medicine*. Baltimore: Williams & Wilkins, 1939.

Atkinson, Rita L., et al. *Introduction to Psychology*. 9th ed. San Diego: Harcourt, Brace, Jovanovich, 1987.

Bailey, Pearce, et al. *The Medical Department of the United States Army in the World War*. Vol. 10, *Neuropsychiatry*. Washington, D.C.: US Government Printing Office, 1929.

Balfour, Harold. *Wings over Westminster*. London: Hutchinson, 1973.

Barris, Ted. *Behind the Glory*. Toronto: Macmillan, 1992.

Bartlett, F.C. *Psychology and the Soldier*. London: Cambridge University Press, 1927.

Bercuson, David J., and S.F. Wise. *The Valour and the Horror Revisited*. Montreal and Kingston, Ont.: McGill-Queen's University Press, 1994.

Bezeau, M.V. "DHist Interim Report on the Computer Study of Second World War RCAF Personnel." Unpublished paper, DHIST, 1 October 1980.

Bingham, Hiram. *An Explorer in the Air Service*. New Haven: Yale University Press, 1920.

Birley, J.L. "A Lecture on the Psychology of Courage. Delivered at the RAF Staff College on March 1st, 1923." *Lancet*, 21 April 1923, 779–85.

– "The Principles of Medical Science as Applied to Military Aviation, Lecture I." *Lancet*, 29 May 1920, 1147–51.

– "The Principles of Medical Science as Applied to Military Aviation, Lecture II." *Lancet*, 5 June 1920, 1205–11.

– "The Principles of Medical Science as Applied to Military Aviation, Lecture III." *Lancet*, 12 June 1920, 1251–7.

Bomber Command Quarterly Review (January–March 1944).

Bott, E.A. "Memorandum in [sic] Personnel Selection in the R.A.F." FPRC Report 362, 1941 (PRO, AIR 57).

– "Report by the Subcommittee on Personnel Selection Associate Committee on Aviation Medical Research, Jan to Mar 1941." NRC Report C-2025, 1941.

Bott, E.A., and G.P. Cosgrave. "Preliminary Studies in Selection, R.C.A.F." NRC Report C-2946, August 1941.

Bowyer, Michael J.F. *2 Group RAF*. London: Faber and Faber, 1974.

Boyle, Andrew. *Trenchard*. London: Collins, 1962.

Broad, W., and N. Wade. *Betrayers of the Truth*. New York: Simon & Schuster, 1982.

Broadfoot, Barry. *Six War Years 1939–1945: Memories of Canadians at Home and Abroad*. Toronto: Doubleday, 1974.

Brophy, John, and Eric Partridge. *The Long Trail*. London: Sphere Books, 1969.

Bruce, Herbert A. *Politics and the Canadian Army Medical Corps*. Toronto: William Briggs, 1919.

Buckle, D.F. "The Prevention of Psychiatric Disorders in Flying Personnel." *Medical Journal of Australia* 2, 14 August 1942, 124–6.

Burns, E.L.M. *Manpower in the Canadian Army 1939–1945*. Toronto: Clarke, Irwin, 1956.

Carter, William S. *Anglo-Canadian Wartime Relations, 1939–1945, RAF Bomber Command and No. 6 [Canadian] Group*. Modern European History – Great Britain, ed. Peter Stansky. New York: Garland Publishing, 1991.

Carver, H.S.M. "Personnel Selection in the Canadian Army." Unpublished paper, Directorate of Personnel Selection, National Defence Headquarters (Ottawa), 1945.

Catano, V.M. "Canadian Automated Pilot Selection System: Methods for Assessing Utility." CFPARU Working Paper 90–4, May 1990.

Chajkowsky, William E. *The History of Camp Borden, 1916–18*. Jordan Station, Ont.: Station Press, 1983.

– *Royal Flying Corps: Borden to Texas to Beamsville*. Cheltenham, Ont.: Boston Mills Press, 1979.

Chalke, F.C.R. "Psychiatric Screening of Recruits." *Department of Veterans Affairs Treatment Services Bulletin* 9 (January–June 1954): 273–92.

Chant, S.N.F. "Psychology as Applied in the Royal Canadian Air Force." BCPA 3 (1943): 34–7.

Chappelow, John. "Human Factors in Aviation." *Air Clues* 45 (June 1991): 222–3.

Collier, Basil. *Heavenly Adventurer: Sefton Brancker and the Dawn of British Aviation*. London: Secker & Warburg, 1959.

Conrad, Peter C. *Training for Victory: The British Commonwealth Air Training Plan in the West*. Saskatoon: Western Producer Prairie Books, 1989.

Contemporary Civilization Staff, Columbia College. *Introduction to Contemporary Civilization in the West*. New York: Columbia University Press, 1960.

Copp, Terry. "From Flanders Fields to Sarajevo: Battle Exhaustion and the Canadian Army." Paper presented at the Royal Military College of Canada, Kingston, Ont., 8 March 1993.

Copp, Terry, and Bill McAndrew. *Battle Exhaustion: Soldiers and Psychiatrists in the Canadian Army, 1939–1945*. Montreal and Kingston, Ont.: McGill-Queen's University Press, 1990.

Corner, H.W. "Flying Duties at a Fighter Squadron." FPRC Report 122, 24 March 1940 (PRO, AIR 57).

Dearnaley, E.J., and P.B. Warr, eds. *Aircrew Stress in Wartime Operations*. London: Academic Press, 1979.

Department of National Defence (Ottawa). Defence Research Board Operational Research Group. "A Follow-up Study of Initial Psychiatric Appraisal of Wartime RCAF Aircrew." Operational Research Group Report No. 2, October 1950.

Diringshofen, Heinz von. *Medical Guide for Flying Personnel*. Trans. by Velyien E. Henderson. Toronto: University of Toronto Press, 1940.

D.M.S. (Air). "Factors in Aircrew Selection. Special Study No. 2. The Link Test of Pilot Aptitude." NRC Report C-2787, 17 February 1943.

Dockeray, F.C., and S. Isaacs. "Psychological Research in Aviation in Italy, France, England, and the American Expeditionary Forces." *Journal of Comparative Psychology* 1 (1921): 115–48.

Douglas, W.A.B. *The Creation of a National Air Force*. Vol. 2 of *The Official History of the Royal Canadian Air Force*. Toronto: University of Toronto Press, 1986.

Downes, Cathy. *Special Trust and Confidence: The Making of an Officer*. London: Frank Cass, 1991.

Dreisziger, N.F., ed. *Mobilization for Total War*. Waterloo, Ont.: Wilfrid Laurier University Press, 1981.

Dudley, Sheldon F. "Active Service Flying: The Medical Point of View." *Journal of the Royal Naval Medical Service* 4 (April 1918): 131–40.

Dunlap, Knight. "Psychological Research in Aviation." *Science* 49, 17 January 1917, 1392–3.

Dunmore, Spencer, and William Carter. *Reap the Whirlwind*. Toronto: McClelland & Stewart, 1991.

Ellis, John. *The Social History of the Machine Gun*. New York: Pantheon Books, 1975.

English, Allan D. "Canadian Psychologists and the Aerodrome of Democracy." *Canadian Psychology* 33 (October 1992): 663–72.

English, Allan D., and Michael Rodgers. "Déjà Vu? Cultural Influences on Aviator Selection." *Military Psychology* 4 (1992): 35–47.

English, John A. *The Canadian Army and the Normandy Campaign*. New York: Praeger, 1991.

– *On Infantry*. New York: Praeger, 1984.

Ettinger, G.H. *History of the Associate Committee on Medical Research*. Ottawa: NRC (1946).

Fancher, R.E. *The Intelligence Men*. New York: W.W. Norton, 1985.

– *Pioneers of Psychology*. 2nd ed. New York: W.W. Norton, 1990.

Feasby, W.R., ed. *The Official History of the Canadian Medical Services 1939–1945*. 2 vols. Ottawa: Queen's Printer, 1953.

Fernandez-Armesto, Lesley. "The Experiences of Officers in a New Service: The RAF 1918–1939." Unpublished paper, King's College, London, 1980 (copy in RAFSC Library, Bracknell).

Foster, J.A. *For Love and Glory*. Toronto: McClelland & Stewart, 1989.

Fredette, Raymond H. *The Sky on Fire*. New York: Holt, 1966.

Fussell, Paul. *Wartime: Understanding and Behavior in the Second World War*. New York: Oxford University Press, 1989.

Gal, Reuven, and A. David Mangelsdorff, eds. *Handbook of Military Psychology*. Chichester: John Wiley, 1991.

Gibson, T.M., and M.H. Harrison. *Into Thin Air: A History of Aviation Medicine in the RAF*. London: Robert Hale, 1984.

Gilchrist, Norman S. "An Analysis of Causes of Breakdown in Flying." *British Medical Journal* 2, 12 October 1918, 401–3.

Gillespie, R.D. *Psychological Effects of War on Citizen and Soldier*. New York: W.W. Norton, 1942.

– "Report on Visits to Stations in Coastal Command to Investigate Neuropsychiatric Problems." FPRC Report 86b, 8 October 1939 (PRO, AIR 57).

Gould, Stephen Jay. *The Mismeasure of Man*. New York: W.W. Norton, 1981.

Granatstein, J.L. *Canada's War: The Politics of the Mackenzie King Government 1939–1945*. Toronto: Oxford University Press, 1975.

Granatstein, J.L., and J.M. Hitsman. *Broken Promises: A History of Conscription in Canada*. Toronto: Oxford University Press, 1977.

Granatstein, J.L., and Desmond Morton. *A Nation Forged in Fire*. Toronto: Lester & Orpen Dennys, 1989.

Greenhous, Brereton, ed. *A Rattle of Pebbles: The First World War Diaries of Two Canadian Airmen*. Ottawa: Canadian Government Publishing Centre, 1987.

Greenhous, Brereton, et al. *The Crucible of War, 1939–45*. Vol. 3 of *The Official History of the Royal Canadian Air Force*. Toronto: University of Toronto Press, 1994.

G[rey], C.G. "A Question of Ethnology." *The Aeroplane* 3, 28 August 1913, 227–8.

Grindley, G.C. "Note on 'Flying Stress' after Reconnaissance Flights." FPRC Report 121, [February (?) 1940] (PRO, AIR 57).

Grinker, Roy R., and John P. Spiegel. *Men under Stress*. Philadelphia: Blakiston, 1945.

Hamel, Gustav, and Charles C. Turner. *Flying: Some Practical Aspects*. London: Longmans, Green, 1914.

Harris, Arthur. *Bomber Offensive*. London: Collins, 1947.

Harvey, J. Douglas. *Boys, Bombs and Brussels Sprouts*. Toronto: McClelland & Stewart, 1981.
– *Laughter-Silvered Wings*. Toronto: McClelland & Stewart, 1984.
– *The Tumbling Mirth*. Toronto: McClelland & Stewart, 1983.
Hastings, Donald W.; David G. Wright; and Bernard C. Glueck. "Psychiatric Experiences of the Eighth Air Force: First Year of Combat (July 4, 1942–July 4, 1943)." U.S. Army Air Forces, Josiah Macy, Jr. Foundation, (1943).
Hastings, Max. *Bomber Command*. New York: Dial Press, 1979.
Hatch, F.J. *Aerodrome of Democracy*. Ottawa: DHIST, 1983.
Hawker, Tyrrel Mann. *Hawker, V.C.* London: Mitre Press, 1965.
Haycock, Ronald G. *Sam Hughes* (Canadian War Museum Historical Publication No. 21). Waterloo, Ont.: Wilfrid Laurier University Press, 1986.
Hayes, Geoffrey. "Science and the Magic Eye: Innovations in the Selection of Canadian Army Officers 1939–1945." Paper presented at the Annual Meeting of the Society for Military History, Royal Military College of Canada, Kingston, Ont., 21 May 1993.
Henmon, V.A.C. "Air Service Tests of Aptitude for Flying." *Journal of Applied Psychology* 3 (1919): 103–9.
Higham, Robin. *A Guide to the Sources of British Military History*. Berkeley: University of California Press, 1971.
Hill, Bradford, and Denis Williams. "Investigation into Psychological Disorders in Flying Personnel: The Reliability of Psychiatric Opinion in the RAF." FPRC Report 601, October 1944 (PRO, AIR 57).
Hitchins, F.H. *Air Board, Canadian Air Force and Royal Canadian Air Force* (Mercury Series, Canadian War Museum Paper No. 2). Ottawa: Canadian War Museum, 1972.
Hitsman, J.M. "Manpower Problems of the RCAF during the Second World War." Army HQ Report No. 67, 15 January 1954 (copy at DHIST).
Hunt, B.D., and R.G. Haycock, eds. *Canada's Defence: Perspectives on Policy in the Twentieth Century*. Toronto: Copp Clark Pitman, 1993.
Hyatt, A.M.J. *General Sir Arthur Currie: A Military Biography*. Toronto: University of Toronto Press, 1987.
Insall, A.J. *Observer: Memories of the RFC 1915–1918*. London: William Kimber, 1970.
Ironside, R.N., and I.R.C. Batchelor. *Aviation Neuro-Psychiatry*. Edinburgh: E & S Livingstone, 1945.
Janis, Irving L. *Air War and Emotional Stress*. New York: McGraw-Hill, 1951.
Jensen, Richard S., ed. *Aviation Psychology*. Aldershot: Gower Technical, 1989.
Johnston, Neil; Ray Fuller; and Nick McDonald,. eds. *Aviation Psychology: Training and Selection*. Aldershot: Avebury Aviation, 1995.

Jones, H.A. *War in the Air*. See Raleigh, Walter.

Jones, Isaac. *Flying Vistas*. Philadelphia: Lippincott, 1937.

Jordan, Gerald, ed. *British Military History: A Supplement to Robin Higham's Guide to the Sources*. New York: Garland Publishing, 1988.

Kellet, Anthony. *Combat Motivation*. Boston: Kluwer, 1982.

Kennett, Lee. *The First Air War*. New York: The Free Press, 1991.

Kingston-McCloughry, E.J. "Leadership with Special Reference to World War II." *RAF Quarterly* 7, no. 3 (Autumn 1967): 201–9.

Kinney, Curtis, with Dale M. Titler. *I Flew a Camel*. Philadelphia: Dorrance, 1972.

Kitchen, Peter A. "Canada's Air Force – Developing a New Flight Plan." *Forum: Journal of the Conference of Defence Associations Institute* 7, no. 3 (September 1992): 3.

L'Etang, H.J.C.J. "A Criticism of Military Psychiatry in the Second World War." *Journal of the Royal Army Medical Corps* 97 (1951): 192–7.

Lewis, Bruce. *Aircrew: The Story of the Men Who Flew the Bombers*. London: Leo Cooper, 1991.

Lewis, Cecil. *Sagittarius Rising*. London: Peter Davies, 1936.

Lewis, Gwilym H. *Wings over the Somme*. Ed. by Chaz Bowyer. London: William Kimber, 1976.

Lightman, Ernie Stanley. "The Economics of Military Manpower Supply in Canada." PhD diss., University of California, Berkeley, 1972.

Livingston, Philip. *Fringe of the Clouds*. Toronto: Ryerson Press, 1962.

Longmate, Norman. *The Bombers*. London: Hutchinson, 1983.

Loo, Robert. "Pilot Selection: Past, Present and Future." Paper presented to the 25th Human Factors Conference, Rochester, N.Y., 12–16 October 1981.

McAndrew, William. "From Integration to Separation: The RCAF's Evolution to Independence." *Revue Internationale d'Histoire Militaire* 54 (1982): 131–58.

McCarthy, John. "Aircrew and 'Lack of Moral Fibre' in the Second World War." *War and Society* 2, no. 3 (September 1984): 87–101.

McCudden, James Byford. *Flying Fury*. London: Aviation Book Club, 1930.

MacCurdy, J.T. *War Neuroses*. London: Cambridge University Press, 1918.

McDonald, Nick; Neil Johnston; and Ray Fuller, eds. *Applications of Psychology to the Aviation System*. Aldershot: Avebury Aviation, 1995.

McIntosh, Dave. *Terror in the Starboard Seat*. Don Mills, Ont.: General Publishing, 1980.

McKay, Russell. *One of the Many*. Burnstown, Ont.: General Store Publishing House, 1989.

Macmillan, Norman. *Into the Blue*. London: Duckworth, 1929.

– *Offensive Patrol*. London: Jarrolds, 1973.

MacNalty, Arthur Salusbury, and W. Franklin Mellor, eds. *Medical Services in War*. London: HMSO, 1968.

McNeill, William H. *The Pursuit of Power: Technology, Armed Force, and Society since A.D. 1000*. Chicago: University of Chicago Press, 1982.

Macphail, Andrew. *The Medical Services: Official History of the Canadian Forces in the Great War 1914–19*. Ottawa: King's Printer, 1925.

Macpherson, W.G. *History of the Great War – Medical Services: General History*. 2 vols. London: HMSO, 1921.

Mashburn, Neely C. "Some Interesting Psychological Factors in the Selection of Military Aviators." *Journal of Aviation Medicine* 6 (1935): 113–26.

Mason, R.A. *History of the RAF Staff College 1922–72*. Bracknell, RAFSC, 1972.

Mauer, Mauer. *Aviation in the U.S. Army, 1919–1939*. Washington, D.C.: Office of Air Force History, 1987.

919–1939. Washington, D.C.: Office of Air Force History, 1987.

Maycock, Robert. *Doctors in the Air*. London: George Allen & Unwin, 1957.

"The Medial Service of the Air Force." *Lancet*, 16 March 1918, 419.

Messenger, Charles. *"Bomber" Harris and the Strategic Bombing Offensive, 1939–1945*. London: Arms and Armour Press, 1984.

Middlebrook, Martin, and Chris Everitt. *The Bomber Command War Diaries*. London: Penguin, 1990.

Milberry, Larry. *Aviation in Canada*. Toronto: McGraw-Hill Ryerson, 1979.

Milner, Marc, ed. *Canadian Military History*. Toronto: Copp Clark Pitman, 1993.

Mitchell, T.J., and G.M. Smith. *History of the Great War – Medical Services: Casualties and Medical Statistics*. London: HMSO, 1931.

Moran, Lord. *The Anatomy of Courage*. 2nd ed. London: Constable, 1966.

Morgan, Kenneth O., ed. *The Oxford Illustrated History of Britain*. Oxford: Oxford University Press, 1984.

Morton, Desmond. *Canada and War*. Political Issues in Their Historical Perspectives, eds. G.A. Rawlyk and B.W. Hodgins. Toronto: Butterworths, 1981.

– *A Military History of Canada*. Edmonton: Hurtig Publishers, 1985.

Murray, Williamson. *Strategy for Defeat*. Secaucus, N.J.: Chartwell, 1986.

Myers, C.R. "Edward Alexander Bott." *Canadian Psychologist* 3 (1974): 292–302.

National Research Council of Canada. *History of the Associate Committee on Aviation Medical Research 1939–1945*. Ottawa: NRC, 1946.

"The New Certificate Tests." *The Aeroplane* 3, 28 August 1913, 227.

"News and Comment." *BCPA* 1 (October 1940): 6.

"News and Comment." *BCPA* 6 (February 1946): 15.

Paterson, T.T. *Morale in War and Work*. London: Max Parrish, 1955.

Pavlov Institute of Aviation Medicine. *Fundamentals of Aviation Medicine*. Trans. by Iser Steiman. Ottawa: NRC, 1943.

Peden, Murray. *A Thousand Shall Fall*. Stittsville, Ont.: Canada's Wings, 1979.

Plumptre, A.F.W. *Mobilizing Canada's Resources for War*. Toronto: Macmillan, 1941.

Privy Council Office (United Kingdom). "Report of an Expert Committee on the Work of Psychologists and Psychiatrists in the Services." London: HMSO, 1947.

Rachman, Stanley J. *Fear and Courage*. San Francisco: Freeman, 1978.

Raleigh, Walter, and H.A. Jones. *War in the Air*. 6 vols. and appendixes. Oxford: Clarendon Press, 1922–37.

Rees, J.R. "Three Years of Military Psychiatry in the United Kingdom." *British Medical Journal* 1, 2 January 1943, 1–6.

Reid, D.D. "The Operational Aspect of the Medical Supervision of Flying Personnel," 15 December 1941 (PRO, AIR 14/2821).

– "Prognosis for a Return to Full Flying Duties after Psychological Disorder." FPRC Report 535, June 1943. Reprinted in Air Ministry, *Psychological Disorders in Flying Personnel of the RAF*, 234–44.

– "A Study of Some Factors in the Causation of Flying Stress." FPRC Report 450, April 1942 (PRO, AIR 57).

Rexford-Welch, S.C., ed. *The Royal Air Force Medical Services*. 2 vols. London: HMSO, 1954–55.

Rippon, T.S., and E.G. Manuel. "The Essential Characteristics of Successful and Unsuccessful Aviators." *Lancet*, 28 September 1918, 411–15.

Roberts, Leslie. *There Shall Be Wings*. Toronto: Clarke Irwin, 1959.

Robinson, Douglas H. *The Dangerous Sky: A History of Aviation Medicine*. Seattle: University of Washington Press, 1973.

Ruck, Calvin W. "Blacks – They Also Served King and Country." *Maritime Command Trident*, 7 November 1990, 11.

Ruggle, Richard E. Review of Terry Copp and Bill McAndrew, *Battle Exhaustion: Soldiers and Psychiatrists in the Canadian Army, 1939–1945*. *Canadian Defence Quarterly* 20, no. 5 (Spring 1991): 46.

Samelson, Franz. "World War I Intelligence Testing and the Development of Psychology." *Journal of the History of the Behavioral Sciences* 13 (1977): 274–82.

Saward, Dudley. *"Bomber" Harris*. London: Cassell, 1984.

Schwartz, Barry. "The Creation and Destruction of Value." *American Psychologist* 45 (January 1990): 7–15.

Shalit, Ben. *Psychology of Conflict and Combat*. New York: Praeger, 1988.

Signori, E.I. "The Arnprior Experiment: A Study of World War II Pilot Selection Procedures in the RCAF and RAF." *Canadian Journal of Psychology* 3 (1949): 136–50.

Smith, Wilfred I. *Code Word CANLOAN*. Toronto: Dundurn Press, 1992.

Spinner, Barry. "Using the Canadian Automated Pilot Selection System to Predict Performance in Primary Flying School: Derivation and Cross-Validation." CFPARU Working Paper 89–8, 1989.

Stacey, C.P. *Arms, Men and Governments*. Ottawa: Queen's Printer, 1970.

– *A Date with History*. Ottawa: Deneau, (1983).

– *Six Years of War*. Vol. 1 of *The Official History of the Canadian Army in the Second World War*. Ottawa: Queen's Printer, 1955.

– *A Very Double Life: The Private World of Mackenzie King*. Toronto: Macmillan, 1976.

Staff Psychological Section, Office of the Surgeon, Army Air Forces Training Command. "Psychological Activities in the Training Command, Army Air Forces." *Psychological Bulletin* 42 (1945): 37–54.

Stafford-Clark, D. "Morale and Flying Experience: Results of a Wartime Study." *Journal of Mental Science* 95 (1949): 10–50.

Staffs, Psychological Research Unit No. 2 and Department of Psychology, Research Section, School of Aviation Medicine. "Research Program on Psychomotor Tests in the Army Air Forces." *Psychological Bulletin* 41 (1944): 307–21.

Stokes, Alan, and Kirsten Kite. *Flight Stress*. Aldershot: Avebury Aviation, 1994.

Stradling, A.H. *Customs of the Service*. Aldershot: Wellington Press, 1947.

Strange, L.A. *Recollections of an Airman*. London: Greenhill Books, 1989.

Stratton, George M., et al. "Psychological Tests for Selecting Aviators." *Journal of Experimental Psychology* 6 (December 1920): 405–23.

Sullivan, Alan. *Aviation in Canada 1917–18*. Toronto: Rous & Mann, 1919.

Sutton, D.G. "Psychology in Aviation." *U.S. Naval Medical Bulletin* 28 (1930), 5–13.

Swettenham, John. *McNaughton*. Vol. 1: *1887–1939*. Toronto: Ryerson Press, 1968.

Symonds, C.P. "The Human Response to Flying Stress." Reprinted in Air Ministry, *Psychological Disorders in Flying Personnel of the RAF*, 100–16.

– "Memorandum on the Use and Abuse of the Term 'Flying Stress.'" FPRC Report 412 (1941) (PRO, AIR 57).

– "Report on Visits to RAF Hospital Cranwell and to Operational Stations in Fighter Command to Investigate Neuropsychiatric Problems." FPRC Report 86c, 8 September 1939 (PRO, AIR 57).

– *Studies in Neurology*. New York: Oxford University Press, 1970.

Symonds, C.P., and Denis Williams. "Clinical and Statistical Study of Neurosis Precipitated by Flying Duties." FPRC Report 547, August 1943. Reprinted in Air Ministry, *Psychological Disorders in Flying Personnel of the RAF*, 140–72.

– "Investigation of Psychological Disorders in Flying Personnel by Unit Medical Officers." FPRC Report 412(k), February 1945 (PRO, AIR 57).

– "Occurrence of Neurosis in Royal Air Force Air Crew in 1943 and 1944." FPRC Report 412(i), September 1944. Reprinted in Air Ministry, *Psychological Disorders in Flying Personnel of the RAF*, 173–7.

– "Occurrence of Neurosis in Royal Air Force Air Crews in 1944 and 1945." FPRC Report 412(L), April 1945. Reprinted in Air Ministry, *Psychological Disorders in Flying Personnel of the RAF*, 178–82.

– "Personal Investigation of Psychological Disorders in Flying Personnel of Bomber Command." FPRC Report 412(f), August 1942. Reprinted in Air Ministry, *Psychological Disorders in Flying Personnel of the RAF*, 31–64.

– "The Probability of Return to Full Flying of Men Who Have Broken Down under the Strain of Operational Duties." FPRC Report 561, November 1943 (PRO, AIR 57).

– "Psychological Disorders in Flying Personnel, Section 1. A Critical Review of Published Literature." FPRC Report 412(c), January 1942 (PRO, AIR 57).

– "Psychological Disorders in Flying Personnel, Section 2. Review of Reports Submitted to Air Ministry since the Outbreak of the War, April 1942." FPRC Report 412(d) (1942) (PRO, AIR 57).

– "Statistical Survey of the Occurrence of Psychological Disorders in Flying Personnel in the Six Months February to August 1942." FPRC Report 412(g), November 1942. Reprinted in Air Ministry, *Psychological Disorders in Flying Personnel of the RAF*, 117–39.

Taylor, Gordon. *Sopwith Scout 7309*. London: Cassell, 1968.

Terraine, John. *The Right of the Line*. London: Hodder and Stoughton, 1985.

Thompson, Walter. *Lancaster to Berlin*. Toronto: Totem Books, 1987.

Thorndike, Edward L. "Scientific Personnel Work in the Army." *Science* 49, 17 January 1919, 53–61.

– "The Selection of Military Aviators – Mental and Moral Qualities." *US Air Service Journal* 1 (June 1919): 14–17.

– "The Selection of Military Aviators – III – The Selective Action of the Ground Schools." *US Air Service Journal* 2 (January 1920): 29–31.

Van Creveld, Martin. *Command in War*. Cambridge, Mass.: Harvard University Press, 1985.

– *Supplying War*. Cambridge, England: Cambridge University Press, 1977.

– *Technology and War*. New York: The Free Press, 1991.

Vee, Roger [Vivian Voss]. *Flying Minnows: Memoirs of a World War I Fighter Pilot from Training in Canada to the Front Line, 1917–1918*. London: Arms and Armour Press, 1976.

Vernon, P.E., and J.B. Parry. *Personnel Selection in the British Forces*. London: University of London Press, 1949.

Verrier, Anthony. *The Bomber Offensive*. 2nd ed. London: Pan Books, 1974.

Villars, Jean Beraud. *Notes of a Lost Pilot*. Trans. by Stanley J. Pincetl, Jr., and Ernest Marchand. Hamden, Conn.: Archon Books, 1975.

Walker, R.J. "Poles Apart: Civil-Military Relations in the Pursuit of a Canadian National Army." MA thesis, Royal Military College of Canada, 1991.

Ward, Norman, ed. *A Party Politician: The Memoirs of Chubby Power*. Toronto: Macmillan, 1966.

Webster, Charles, and Noble Frankland. *The Strategic Air Offensive against Germany 1939–1945*. 4 vols. London: HMSO, 1961.

Weisbord, Merrily, and Merilyn Simonds Mohr. *The Valour and the Horror*. Toronto: Harper Collins, 1991.

Wilbur, F.I. "Aviation and Common Sense." *Flight* 3, 6 May 1911, 399–400.

Williams, D.C. "The Frustrating Fifties." Address to the 50th Anniversary Meeting of the Canadian Psychological Association, Halifax, N.S., June 1989.

– "Sperrin Noah Fulton Chant." *Canadian Psychology* 32 (1991): 86.

Williams, Denis J. "A Note on the Recognition of Neurosis in Flying Personnel." FPRC Report 412(j), February 1944 (PRO, AIR 57).

– "Predisposition to Psychological Disorder in Normal Flying Personnel." FPRC Report 516, February 1943. Reprinted in Air Ministry, *Psychological Disorders in Flying Personnel of the RAF*, 185–92.

Williams, James N. *The Plan: Memories of the BCATP*. Stittsville, Ont.: Canada's Wings, 1984.

Wilson, Fred, and Karol Wenek. "A Clear Conscience: Psychology in the Canadian Military and Social Responsibility." *Psynopsis* (Spring 1993): 16–17.

Winter, Denis. *The First of the Few: Fighter Pilots of the First World War*. Athens, Ga.: University of Georgia Press, 1983.

Wise, S.F. *Canadian Airmen and the First World War*. Vol. 1 of *The Official History of the Royal Canadian Air Force*. Toronto: University of Toronto Press, 1980.

Wiskoff, M.F., and G.M. Rampton, eds. *Military Personnel Measurement*. New York: Praeger, 1989.

Wortley, Rothesay Stuart. *Letters from a Flying Officer*. London: Oxford University Press, 1929.

Wright, M.J. "CPA: The First Ten Years." *Canadian Psychologist* 2 (1974), 112–31.

Wright, M.J., and C.R. Myers, eds. *History of Academic Psychology in Canada*. Toronto: C.J. Hogrefe, 1982.

Index